Praise for This Book

I am impressed that the authors cover so many Rasch measurement topics perfectly in this small book. Rasch-related measurement theories and fundamental concepts with illustrative analyses in each chapter make this book an extremely useful reference for Rasch researchers and practitioners. Tables and figures created by the authors are extremely helpful in better understanding the material. This will be required reading in my Rasch models course and I'll recommend it to my graduate students and colleagues who are interested in applying Rasch models.

—Yi-Hsin Chen, *University of South Florida*

This book provides an excellent overview of the theory and special procedures that are available for Rasch model applications.

—Susan Embretson, *Georgia Institute of Technology*

This book is a must-read for anyone who wants to understand how Rasch measurement is much more than a set of statistical models. It is a measurement theory that can help practitioners to achieve their core goals in developing and using scales, and this book details the topics that support such uses of the theory.

—A. Corinne Huggins-Manley, *University of Florida*

This book provides a concise and clear treatment of using Rasch measurement theory in developing and maintaining a scale in the social sciences. The use of a construct map makes for an intuitive understanding of key concepts and procedures such as scale development and measurement invariance.

—Yanyan Sheng, *University of Chicago*

To my wonderful family: Judy Monsaas, Emily Monsaas Engelhard, David Monsaas Engelhard, Shimby McCreery, Dashiell Engelhard McCreery, and Cassio Engelhard McCreery. All of you make my life a joy! (GE)
To my husband, Jun Zhang, with love. (JW).

Rasch Models for Solving Measurement Problems

Quantitative Applications in the Social Sciences

A Sage Publications Series

Quantitative Applications in the Social Sciences

A Sage Publications Series

Rasch Models for Solving Measurement Problems

Invariant Measurement in the Social Sciences

George Engelhard, Jr.

University of Georgia

Jue Wang

University of Miami

Los Angeles | London | New Delhi
Singapore | Washington DC | Melbourne

For information:

SAGE Publications, Inc.

2455 Teller Road

Thousand Oaks, California 91320

E-mail: order@sagepub.com

SAGE Publications Ltd.

1 Oliver's Yard

55 City Road

London, EC1Y 1SP

United Kingdom

SAGE Publications India Pvt. Ltd.

B 1/I 1 Mohan Cooperative Industrial Area

Mathura Road, New Delhi 110 044

India

SAGE Publications Asia-Pacific Pte. Ltd.

18 Cross Street #10-10/11/12

China Square Central

Singapore 048423

Printed in the United States of America

ISBN 978-1-5443-6302-8

This book is printed on acid-free paper.

SUSTAINABLE FORESTRY INITIATIVE

Certified Chain of Custody
At Least 10% Certified Forest Content
www.sfiprogram.org
SFI-01028

Acquisitions Editor: Helen Salmon

Editorial Assistant: Elizabeth Cruz

Production Editor: Olivia Weber-Stenis

Copy Editor: TNQ

Typesetter: TNQ

Indexer: TNQ

Cover Designer: Candice Harman

Marketing Manager: Shari Countryman

20 21 22 23 24 10 9 8 7 6 5 4 3 2 1

CONTENTS

SERIES EDITOR'S INTRODUCTION

I am pleased to introduce *Rasch Models for Solving Measurement Problems: Invariant Measurement in the Social Sciences*, by George Engelhard and Jue Wang. While some researchers view Rasch models within the tradition of item response theory (IRT) more generally, the authors of this volume do not. Professors Engelhard and Wang view the requirements of invariant measurement as a set of hypotheses that must be tested and verified, in contrast to IRT where the focus is on how to best fit and reproduce the data. As the title indicates, invariance of a scale is the focus of the volume.

Invariance across persons and invariance across items are properties of Rasch models obtained when there is good model-data fit. Invariance across persons means that no matter which items are selected, persons with greater ability will tend to do better. Invariance across items means that regardless of ability, there must be a greater chance of success on an easy item than a more difficult one. Invariance may be best explained using the metaphor of a ruler. A ruler measures height the same way no matter how the person is being measured; the relative height of two individuals is the same whether measured in inches or centimeters. A ruler also exemplifies another feature of Rasch models: they are unidimensional.

Professors Engelhard and Wang introduce the Rasch measurement theory step by step, with chapters on scale construction, evaluation, maintenance, and use. Scale construction involves defining the latent variable, identifying items that measure it, creating scoring rules, and describing a Wright map (a visual representation of items ordered based on their relative difficulty). The evaluation of a Rasch scale addresses how well the requirements of the Rasch model are satisfied given a specific data set, with particular attention to measurement invariance and differential item functioning. The chapter on scale maintenance takes up issues around the interchangeability of items as well as establishing comparability across persons. The chapter on scale uses addresses foundational concepts of reliability, validity, and fairness in the context of invariant measurement and in relation to the *Standards for Educational and Psychological Testing*.

Rasch Models for Solving Measurement Problems: Invariant Measurement in the Social Sciences is broadly accessible. It could serve as a supplemental text in a graduate course focused on measurement or test development. It also provides a helpful introduction to Rasch measurement theory for interested practitioners. Points are illustrated and techniques demonstrated through an extended example: the Food Insecurity Experience (FIE) scale. The FIE scale measures *access to food* via eight yes/no questions that have meaning and are interpretable in diverse settings across the United States and around the world. Sample data sets are available online, as is syntax for two computer programs, one called FACETS and an R-based program called ERMA (Everyone's Rasch Measurement Analyzer).

As Professors Engelhard and Wang themselves show, Rasch measurement theory is most often applied in psychology, educational research, and health. No doubt, the volume will be of particular interest to researchers working in these fields. That said, the measurement issues it addresses are fundamental to all social science research and thus relevant to a broad audience of practitioners. It deserves a spot on every social scientist's bookshelf.

Barbara Entwisle
Series Editor

PREFACE

Aims of the Book

A major aim of this book is to introduce current perspectives on Rasch measurement theory with an emphasis on developing Rasch-based scales. Rasch measurement theory represents a paradigm shift in measurement theory away from classical test theory and creates a framework for scaling that can yield invariant measurement. Rasch scale development includes four components: constructing, evaluating, maintaining, and using a scale. These components can contribute to the solution of practical measurement problems in the social, behavioral, and health sciences. Many measurement problems are meaningfully understood through a careful consideration of the requirements of invariant measurement as conceptualized within Rasch measurement theory. Our ultimate goal is to encourage readers to join us in the quest for invariant measurement through the development of Rasch scales in the social sciences.

Overall Structure of the Book

Chapter 1 introduces the basic requirements of invariant measurement. Rasch measurement theory is briefly discussed as a system of measurement that can be used to solve practical measurement problems in the social sciences. The components of scale development for creating Rasch scales support the potential to achieve invariant measurement. Four measurement problems are introduced to illustrate how Rasch measurement theory can be used to address these issues.

Chapter 2 examines the basics of constructing a scale to represent a latent variable of interest to the researcher. The requirements of invariant measurement are used to guide each of the components of scale construction based on Rasch measurement theory. The essential components for constructing a Rasch scale include defining the latent variable (construct) to be measured, creation of an observational design (e.g., items or questions), specifying a set of scoring rules, and applying the Rasch model to the observed data to obtain a Wright map that empirically shows the constructed scale.

Chapter 3 discusses the nuts-and-bolts aspects of evaluating a Rasch scale for a specific set of items and group of persons. Invariant measurement includes item-invariant person measurement and person-invariant item calibration, and it can only be realized when good model-data fit is obtained. Strategies for evaluating a Rasch scale take the theoretical and ideal aspects of invariant measurement into the real world of observed measurement.

Chapter 4 introduces an important and general problem encountered in social science measurement: How do we maintain the psychometric quality of a scale over a variety of conditions? It places an emphasis on the comparability of person scores. It should be stressed that it is the connection between the underlying latent variable and the continuum that is viewed as invariant, but not the particular items or persons used to operationalize the scale. In other words, we are maintaining scales to establish a common metric so that different subsets of items and scales can be adjusted to produce comparable person scores.

Chapter 5 discusses the three foundations of testing based on the *Test Standards* (AERA, APA, & NCME, 2014). The three foundations are the validity, reliability, and fairness of the use of test scores. Researchers create scales to use the scores for a variety of purposes. One of the purposes is to inform policy, and it is common in policy setting to identify cut scores or critical points on the scale that define ordered categories. The examples in educational achievement include the Achievement Levels used in the National Assessment of Educational Progress (NAEP), such as Basic, Proficient, and Advanced. It is very important to critically evaluate whether or not the scale scores can be used to guide policy for assessment and testing.

Finally, Chapter 6 summarizes our major points and highlights future directions for readers who are interested in using Rasch measurement theory.

Audiences for the Book

The intended audiences for this book are the same as other books in SAGE's *Quantitative Applications in the Social Sciences* (QASS) series. These "Little Green Books" are designed to be instructionally relevant for students, instructors, and researchers—our contribution is targeted toward the same audiences.

Learning Tools

All of the Rasch analyses in this book were run using the Facets computer program (Linacre, 2018a). We have also created an R program—Everyone's Rasch Measurement Analyzer (ERMA) that can be used to obtain parameter estimates of the Rasch model and to examine model-data fit. There is a constantly evolving set of R packages that can be used for Rasch analysis, and the reader should search the web for recent packages.

We have also created an online module for Rasch measurement theory (Wang & Engelhard, 2019) that is included as part of the ITEMS modules available from the National Council on Measurement in Education (NCME). The module is freely available through the NCME website. Winsteps.com remains an important resource for current advances in Rasch software including Winsteps and Facets. Rasch Measurement Transactions (www.rasch.org) is another useful website for discussions of issues and advances in Rasch Measurement Theory.

Data Sets

A data set based on the Food Insecurity Experience (FIE) scale is used throughout the entire book. This scale is an eight-item scale with the items coded dichotomously (Yes = 1, No = 0). We selected a sample of 40 people in order to illustrate the concepts in this book.

Online Resources

Facets syntax, R code for ERMA, and sample data sets are available on a website for the book at https://study.sagepub.com/researchmethods/qass/engelhard-rasch-models.

ACKNOWLEDGMENTS

We would like to thank and acknowledge the following reviewers for helpful comments and guidance:

Yi-Hsin Chen, *University of South Florida*

Susan Embretson, *Georgia Institute of Technology*

Ben B. Hansen, *University of Michigan*

A. Corinne Huggins-Manley, *University of Florida*

Yanyan Sheng, *University of Chicago*

ABOUT THE AUTHORS

George Engelhard, Jr., PhD, joined the faculty at the University of Georgia in the fall of 2013. He is Professor Emeritus at Emory University (1985–2013). Professor Engelhard received his PhD in 1985 from the University of Chicago (MESA Program—measurement, evaluation, and statistical analysis). Professor Engelhard is the author of two books: *Invariant measurement with raters and rating scales: Rasch models for rater-mediated assessments* (2018 with Dr. Stefanie A. Wind) and *Invariant measurement: Using Rasch models in the social, behavioral, and health sciences* (2013). He is the co-editor of five books, and he has authored or co-authored over 200 journal articles, book chapters, and monographs. Professor Engelhard was a co-editor of the *Journal of Educational Measurement.* He serves on several national technical advisory committees on educational measurement and policy in several states in the United States. In 2015, he received the first Qiyas Award for Excellence in International Educational Assessment recognizing his contributions to the improvement of educational measurement at the local, national, and international levels. He is a fellow of the American Educational Research Association.

Jue Wang, PhD, is an Assistant Professor in the Research, Measurement & Evaluation Program at the University of Miami. Dr. Wang received her PhD in the Quantitative Methodology (QM) Program under Educational Psychology at the University of Georgia (UGA) in 2018. She also obtained an MS degree in Statistics at UGA. While at UGA, Dr. Wang was awarded the Owen Scott Doctoral Research Scholarship in 2016 recognizing the contribution of her research work on evaluating rater accuracy and perception using Rasch measurement theory. Furthermore, she received the QM Outstanding Student Award in 2018 recognizing her accomplishments in research, teaching, and service during graduate school. Her research focuses on examining rating quality and exploring rater perception in rater-mediated assessments using measurement models, such as a family of Rasch models, unfolding models, and multilevel item response models. She has published in major journals related to measurement including *Educational and Psychological Measurement, Journal of Educational Measurement, Assessing Writing,* and *Psychological Test and Assessment Modeling.*

1

INTRODUCTION

Georg Rasch summarized his basic research on measurement in a book entitled *Probabilistic Models for Some Intelligence and Attainment Tests* (1960/1980). Rasch's research led to the development of a new paradigm for measurement in the social sciences. As pointed out by van der Linden (2016), the first chapter of Rasch's book is mandatory reading for anyone seeking to understand the transition from classical test theory (CTT) to item response theory (IRT). Our book focuses on describing Rasch's theoretical and applied contributions to measurement theory and the use of the Rasch model as a framework for solving a set of important measurement problems. We also stress the continuing relevance of Rasch's contributions to modern measurement for the social, behavioral, and health sciences.

What are the key aspects of Rasch's contributions that provided the basis for a paradigm shift in measurement theory? First of all, Rasch recognized that measurement should focus on individuals. In his words, "present day statistical methods are entirely group-centered, so that there is a real need for developing individual-centered statistics" (Rasch, 1961, p. 321). His solution to this problem led him to propose a set of requirements for specific objectivity in the individual-centered measurement:

- The comparison between two stimuli should be independent of which particular individuals were instrumental for the comparison; and

- it should also be independent of which stimuli within the considered class were or might also have been compared.

- Symmetrically, a comparison between two individuals should be independent of which particular stimuli within the class considered were instrumental for the comparison; and

- it should also be independent of which other individuals were also compared on the same or on some other occasion.

(Rasch, 1961, pp. 331–332)

The first two bullet points suggest that item calibrations (stimuli) should be invariant over the persons that are used to obtain the

1

comparisons: person-invariant calibration of items. The last two bullet points suggest that person measurement should be invariant of the particular items (stimuli) that are used to obtain the comparisons: item-invariant measurement of persons.

Second, Rasch proposed a measurement model for achieving item-invariant measurement of persons and person-invariant calibration of items. This contribution is fundamental because it provides the basis for the separation of items and persons. This separation allows for person scores to be independent of the particular items and for item locations (difficulties) to be estimated independently of the particular persons. In Rasch's research related to IRT, he highlights that the probability of a correct response to an item should be a simple function of item difficulty and person ability.

Another important contribution of Rasch measurement theory is that it provides a philosophical approach stressing that measurement requires a strong set of requirements, and that data may or may not fit the Rasch measurement model. This distinction is essential for achieving invariant measurement with real data. The opportunity to refute the model by evaluating model-data fit is a strong characteristic of Rasch's requirements for specific objectivity and the accomplishment of invariant measurement.

This chapter begins with a brief overview of invariant measurement within three research traditions that can be used for classifying measurement theories. We define the role of invariance in each of these traditions and highlight the importance for invariance in measurement. Second, Rasch measurement theory is introduced as a framework for obtaining invariance in the social sciences. The following section provides an overview of the key components that are essential for scale development from the perspective of Rasch measurement theory. Next, we define several major measurement problems that are discussed in this book including the definition of latent variables, evaluation of differential item functioning, examination of the interchangeability of items for person measurement, and creation of performance standards (cut scores) through standard setting. Finally, an overview is provided of the key topics addressed in this book.

1.1 Invariant Measurement

The scientist seeks measures that will stay put while his back is turned.

(Stevens, 1951, p. 21)

This section provides a brief overview of three research traditions that can be used to organize the major theoretical perspectives that dominate measurement theory: test score, scaling, and structural traditions. We discuss how these traditions relate to invariant measurement. Finally, we introduce how Rasch measurement theory (scaling tradition) is used as a framework for understanding invariant measurement.

Research Traditions in Measurement

In order to understand the evolution of measurement theory and the concept of invariance during the 20th century, it is useful to consider three broad paradigms or research traditions that have dominated the field. The first tradition is a test score tradition that is reflected in CTT. The second tradition is a scaling tradition that is reflected in Rasch measurement theory including IRT in general. The third tradition is a structural tradition that can be represented by structural equation modeling (SEM). There are several key distinctions between measurement theories embedded within each research tradition, and these distinctions influence the perspective on invariance within each tradition. These distinctions are (1) focus of the measurement theory, (2) form of the underlying model, (3) perspective on invariance, (4) overarching goal of measurement theories within each tradition, and (5) visual representations of measurement models. Figure 1.1 summarizes some of the major differences between measurement theories that are embedded within these broader research traditions.

The focus of measurement theories within the test score tradition is on sum scores. The underlying statistical model is linear as shown in the standard decomposition of an observed score (O) into a true score (T) and error score (E): $O = T + E$. The underlying statistical model for observed test scores is linear. The definition of CTT is tautological—the model is assumed to be true by definition (Traub, 1997). Measurement theories in the test score tradition focus on the variance of test scores. The overarching goal is to reduce measurement error and minimize potential sources of construct-irrelevant variance. Overall, the test score tradition focuses on reducing noise which is variant. Visual depictions of measurement theories in the test score tradition include the use of Venn diagrams to illustrate different sources of variance in test scores (Cronbach, Gleser, Nanda, & Rajaratnam, 1972).

Measurement theories within the scaling tradition focus on detailed responses of each person to a carefully selected set of items. Rasch

Figure 1.1 Three Traditions in Measurement Theory

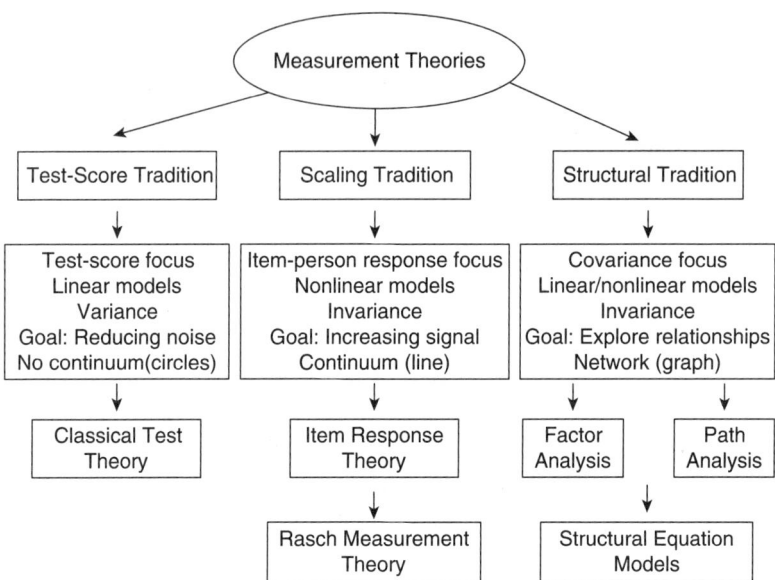

measurement theory as described in this book is an example of models in the scaling tradition. The underlying statistical model is nonlinear and probabilistic. Many of the measurement models in the test score tradition are descriptive in nature, while models within the scaling tradition emphasize strict requirements for model-data fit in order to support the inferences-related invariant measurement. The form of the Rasch model is shown later in this chapter, and a detailed description of model-data fit is provided in Chapter 3. Measurement theories in the scaling tradition emphasize the quest for invariant measurement and the creation of scales to "stay put while our backs are turned" (Stevens, 1951, p. 21). Measurement theories in the scaling tradition have the goal of increasing the signal in our measures, as compared to the reduction of noise emphasized in the test score tradition. The visual depictions of measurement theories in the scaling tradition involve the creation of continuums to represent latent variables (constructs) on a line.

The final family of measurement theories we briefly consider here is within the structural tradition. The structural tradition focuses on covariances (or correlations). Both linear and nonlinear models have been used to examine the relationships between observable and latent variables. Measurement invariance related to models within the structural tradition has been extensively discussed by Millsap (2011). The overarching goal of structural models is to explore relationships between latent variables. Both factor analysis and path analysis including the combination of these models within SEM are multivariate (Bollen, 1989). These networks of variables are typically represented as path diagrams and graphs.

A useful mnemonic device to remember distinctions between the traditions is that the test score tradition relates to circles, the scaling tradition relates to lines, and the structural tradition relates to graphs. The first two traditions were described earlier by Engelhard (2013), while the third tradition is a new addition to the conceptual framework for grouping measurement theories into research traditions. This book focuses on Rasch measurement theory that is within the scaling tradition and the views of invariant measurement from this research tradition.

Invariant Measurement and the Scaling Tradition

Turning now to invariant measurement within the scaling tradition, the basic requirements have been widely recognized by measurement theorists throughout the 20th century. In one of the first books on measurement, Thorndike (1904) outlined the basic ideas of invariant measurement (Engelhard, 2013). It is important to recognize that invariant measurement has two aspects: item-invariant person measurement and person-invariant item calibration. Figure 1.2 lists the requirements for invariant measurement related to Rasch measurement theory. Requirements 1 and 2 imply item-invariant person measurement. Requirements 3 and 4 imply person-invariant item calibration, while the last requirement implies that invariant measures should be unidimensional. Each of these requirements is described in more detail later in our book.

It is of critical importance to recognize that the requirements of invariant measurement are a set of hypotheses that must be verified with a variety of model-data fit procedures. In this book, we stress the use of residual analyses and fit indices for evaluating invariance for person measurement, item calibration, and unidimensionality of the scale. Invariant measurement is not automatically obtained by the

Figure 1.2 Five Requirements for Invariant Measurement

Five Requirements for Invariant Measurement

Item-invariant measurement of persons

1. The measurement of persons must be independent of the particular items that happen to be used for the measuring.

2. A more able person must always have a better chance of success on an item than a less able person (Non-crossing person response functions).

Person-invariant calibration of test items

3. The calibration of the items must be independent of the particular persons used for calibration.

4. Any person must have a better chance of success on an easy item than on a more difficult item (Non-crossing item response functions).

Unidimensional scale

5. Items and persons must be simultaneously located on a single underlying latent variable (Wright map).

---

use of Rasch models—the requirements must be evaluated with specific samples from the target population. Chapter 3 on evaluating a Rasch scale describes these issues in greater detail.

1.2 Rasch Measurement Theory

The psychometric methods introduced in [Rasch's book] ... embody the essential principles of measurement itself, the principles on which objectivity and reproducibility, indeed all scientific knowledge are based.

(Wright, 1980, p. xix)

Rasch measurement theory provides a basis for meeting the requirements of invariant measurement as described in Figure 1.2. The focus of our book is on Rasch measurement theory, but it should be recognized that other measurement models within the scaling tradition, such as IRT, can also meet some of these requirements for invariant measurement. Rasch measurement theory is based on a simple idea about what happens when a single person encounters a single test item (Rasch, 1960/1980). Rasch viewed the probability of a person getting an item correct (or endorsing an item) as a function of the person's ability measure and the difficulty of the item.

Rasch measurement models are fundamentally based on the concept of comparisons. In fact, Rasch (1977) argued that all scientific statements "deal with comparisons, and the comparisons should be objective" (p. 68). Abelson (1995) stressed in his book on *Statistics as Principled Argument* that comparisons are crucial for all meaningful and useful statistical analyses. Within the context of measurement theory, Andrich (1988) highlighted that the "demand for invariance of comparisons across a scale that is unidimensional is paramount" (p. 43). The task for scientists and measurement theorists is to define what is meant by invariant comparisons and objectivity. Rasch proposed a model based on the principle of specific objectivity that provides a framework for these invariant comparisons, namely the Rasch model. The Rasch model for dichotomous responses can be written as:

$$\phi_{ni1} = \frac{\exp(\theta_n - \delta_{i1})}{1 + \exp(\theta_n - \delta_{i1})} \tag{1.1}$$

The Rasch model in Equation 1.1 can be viewed as an operating characteristic function that relates the differences between locations of persons (θ) and items (δ) on a latent variable to the probability of a positive response on a dichotomous item (e.g., correct answer on a multiple-choice item, "yes" response to a survey item). This distance reflects a comparison between each person's ability and item difficulty that predicts a probability of an affirmative answer by a person on each item. This book illustrates the problem-solving capabilities of this elegant approach to measurement proposed by Rasch (1960/1980).

Many current psychometric texts provide general descriptions of IRT models, including the one-parameter logistic (1PL), two-parameter logistic (2PL), and three-parameter logistic (3PL) models (Baker & Kim, 2004). Equation 1.2 shows a general expression for a 3PL model that includes three item parameters—difficulty, discrimination, and pseudo-guessing. The pseudo-guessing parameter defines a lower asymptote for the probability response curve and the minimum probability to get an item correct. If the pseudo-guessing effect is not considered ($c_i = 0$), then this expression reduces to a 2PL model that contains item difficulty and discrimination parameters. Item discrimination parameter reflects the slope or the steepness of an item response function. If a_i is a constant and assuming that item response functions are parallel with common slope, then Equation 1.2 is simplified to represent a 1PL model that estimates only item difficulty parameters. From this perspective, the Rasch model can be obtained when $a_i = 1$ and $c_i = 0$:

$$P_{ij}(X = 1) = c_i + (1 - c_i)\frac{\exp[a_i(\theta_j - b_i)]}{1 + \exp[a_i(\theta_j - b_i)]}, \qquad (1.2)$$

where

P_{ij} = probability of correct response,

b_i = item difficulty parameter,

a_i = item discrimination parameter,

c_i = item lower asymptote (pseudo-guessing) parameter, and

θ_j = latent proficiency of person.

Within the context of general IRT models, some researchers treat the 1PL model as nested within a more general IRT model, and that the 1PL is "equivalent to the well-known Rasch model" (Raykov & Marcoulides, 2011, p. 294). We would like to stress that although the Rasch model can be defined using general IRT model forms, this perspective on the Rasch model does not consider the full philosophical basis of Rasch's contribution as a specific approach to measurement in the social, behavioral, and health sciences. This perspective on the Rasch model as a special case of a more general model creates a specific set of evaluation criteria that stresses how well the models fit data—since including more parameters in a model generally provides better fit, then this perspective, ceteris paribus, privileges more complex IRT models.

In this book, we argue that the fundamental principles and requirements of measurement should come first, and that measurement is the proactive creation of a scale that defines a latent variable based on requirements of invariant measurement—the goal is not to find the best statistical model to fit the data, but the psychometric goal of creating meaningful and useful scales based on a strict set of requirements for high-quality measures. As pointed out by Wright (1980):

If the data cannot be brought into an order which fits with a measurement model, then we will not be able to use these data for measurement, no matter what we do to them. The requirement for results that we are willing to call measures are specified in the model

and if the data cannot be managed by the models, then the data cannot be used to calibrate items or measure persons.

(p. 193)

Andrich (1989) provided an excellent discussion of these issues including an insightful discussion of the distinction between requirements and assumptions in measurement theory.

Rasch measurement theory views model-data fit in terms of a specific set of data meeting the requirements of the model. In order to have the desirable characteristics of invariant measurement, it is essential to have good fit to the model. The basic distinction is that Rasch measurement theory starts with the requirements of the model, while IRT more generally evaluates the success of the models in terms of how well a particular data set is reproduced. We provide a brief comparison between Rasch and 2PL models to show the importance of meeting the requirements of invariant measurement.

The person response functions of four persons are displayed in Figure 1.3 in order to illustrate the importance of item-invariant measurement of persons (Requirements 1 and 2 for invariant measurement). Based on the Rasch model, Persons A, B, C, and D have location measures at -1.00, -0.50, 0.50, and 1.00 logits, respectively. Person A is viewed as having the lowest level of proficiency, while Person D has the highest proficiency on the latent variable. With the 2PL model, slope parameters for persons (Engelhard & Perkins, 2011) are included which are set to 0.30 logits for Person A, 1.60 logits for Person B, 0.80 logits for Person C, and 1.50 logits for Person D.

As we can see in Figure 1.3, the person response functions based on the Rasch model are noncrossing (Panel A); however, the 2PL model yields crossing person response curves (Panel E). Next, probabilities of a correct response are computed for answering three items with difficulty values of -2.00, 0.00, and 2.00 logits separately. Based on the Rasch model, the most proficient Person D has the highest probability of answering each item correct, and Person A who is the least proficient person has the lowest probability of correct on every item. The probability of getting an item correct is higher for Person C who has medium high proficiency than for Person B who has medium low proficiency (Panel B). Based on the 2PL model, the orders of the probabilities are not consistent for different items (Panel F). For instance, the least proficient Person A ($p = 0.29$) had higher probability than Person D ($p = 0.18$) in answering Item 3 correctly. Finally, we order the persons based on their probabilities of answering each item correct.

Figure 1.3 Examining Item-Invariant Measurement of Persons With Person Response Functions

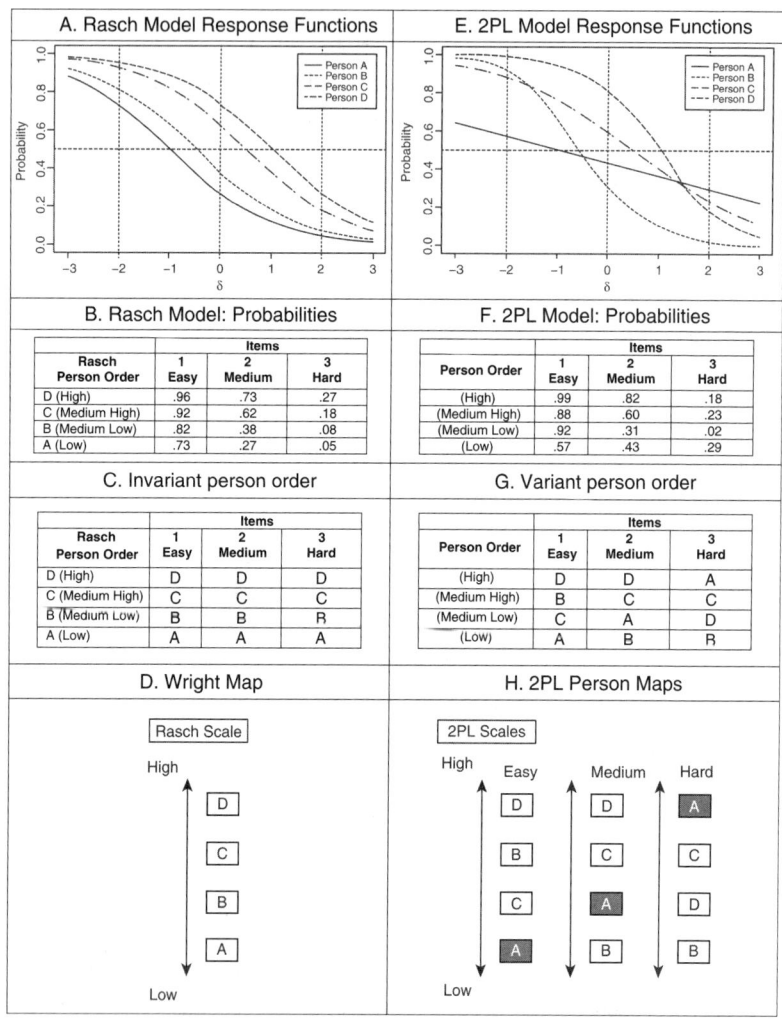

A. Rasch Model Response Functions

E. 2PL Model Response Functions

B. Rasch Model: Probabilities

Rasch Person Order	Items		
	1 Easy	2 Medium	3 Hard
D (High)	.96	.73	.27
C (Medium High)	.92	.62	.18
B (Medium Low)	.82	.38	.08
A (Low)	.73	.27	.05

F. 2PL Model: Probabilities

Person Order	Items		
	1 Easy	2 Medium	3 Hard
(High)	.99	.82	.18
(Medium High)	.88	.60	.23
(Medium Low)	.92	.31	.02
(Low)	.57	.43	.29

C. Invariant person order

Rasch Person Order	Items		
	1 Easy	2 Medium	3 Hard
D (High)	D	D	D
C (Medium High)	C	C	C
B (Medium Low)	B	B	B
A (Low)	A	A	A

G. Variant person order

Person Order	Items		
	1 Easy	2 Medium	3 Hard
(High)	D	D	A
(Medium High)	B	C	C
(Medium Low)	C	A	D
(Low)	A	B	B

D. Wright Map

Rasch Scale

High

D

C

B

A

Low

H. 2PL Person Maps

2PL Scales

High   Easy   Medium   Hard

D   D   A

B   C   C

C   A   D

A   B   B

Low

This explicitly shows the invariant order of persons based on the Rasch model (Panels C and D), but inconsistent patterns obtained with a 2PL model (Panels G and H).

Figure 1.4 illustrates the importance of person-invariant calibration of items (Requirements 3 and 4 for invariant measurement). The probability response functions for three items are shown. The three

Figure 1.4 Examining Person-Invariant Calibration of Items With Item Response Functions

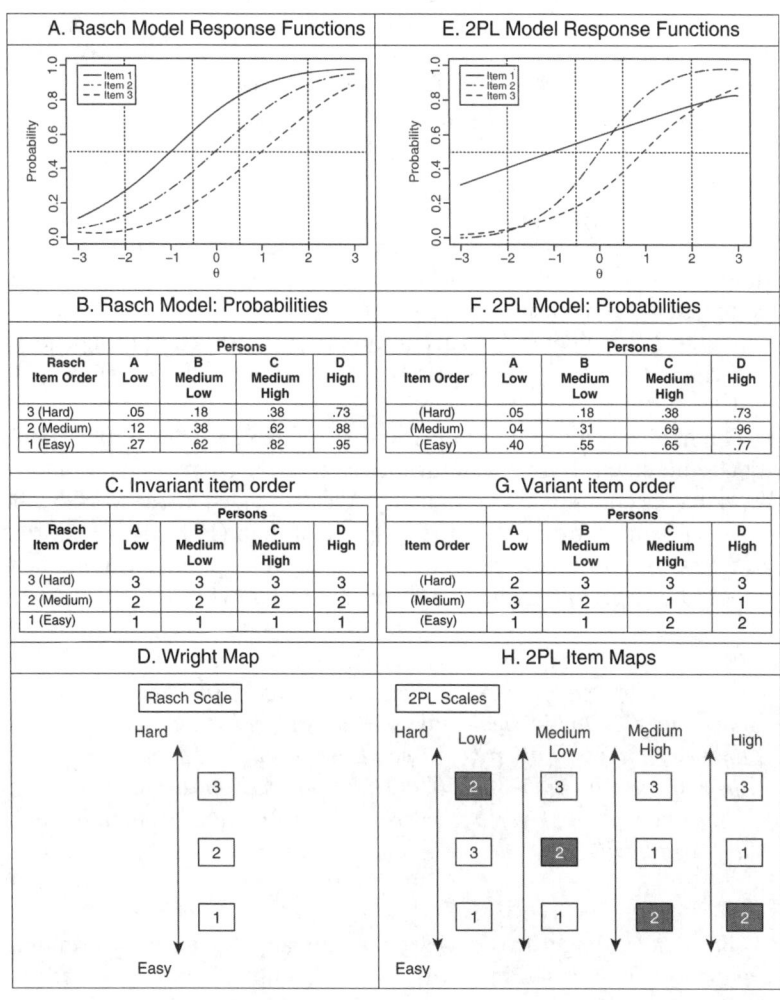

items based on the Rasch model have difficulty values of −1.00 (Item 1), 0.00 (Item 2), and 1.00 (Item 3) logits. Therefore, Item 1 is the easiest, Item 2 is of medium difficulty, and Item 3 is the hardest. The three items of 2PL model not only have the same difficulty parameters as the Rasch items but also have discrimination parameters that vary as follows: 0.40 (Item 1), 1.60 (Item 2), and 1.00 (Item 3) logits.

The Rasch model provides noncrossing item response functions (Panel A); however, the 2PL model has different slopes, and this leads to crossing item response functions (Panel E). Second, the probabilities of getting each item correct have been calculated for four persons—A, B, C, and D with different locations on the latent variable scale which are -2.00, -0.50, 0.50, and 2.00 logits, respectively. Based on a Rasch model, every person has the highest probability of answering the easiest item (Item 1) correct and lowest probability of getting the hardest item (Item 3) correct (Panel B). According to the 2PL model, Person D has higher probability to correctly answer Item 2 ($p = 0.96$), which is of medium difficulty than the easiest Item 1 ($p = 0.77$) as shown in Panel F. Based on the probabilities, we see that the order of the items is invariant with the Rasch model (Panels C and D), while the 2PL model yields variant ordering of items that is dependent on where the person is located on the latent variable (Panels G and H).

The final requirement for invariant measurement is unidimensionality (Requirement 5 for invariant measurement). There are a variety of ways to conceptualize dimensionality. We are guided in our work by the views of Louis Guttman. In the 1940s, Guttman (1944, 1950) laid the groundwork for a new technique designed to explore the unidimensionality of a set of test items. According to Guttman (1950):

> *One of the fundamental problems facing research workers ... is to determine if the questions asked on a given issue have a single meaning for the respondents. Obviously, if a question means different things to different respondents, then there is no way that the respondents can be ranked ... Questions may appear to express a single thought and yet not provide the same kind of stimulus to different people.*
>
> (p. 60)

Guttman scaling can be viewed as an approach for determining whether or not a set of items and a group of persons meet the requirements of unidimensionality (Engelhard, 2008a). As shown in Figures 1.2 and 1.3, the Rasch scale can be viewed as a probabilistic Guttman scale. This is very important as pointed out by Cliff (1983):

> *the Guttman scale is one of the very clearest examples of a good idea in all of psychological measurement. Even with an unsophisticated—but intelligent—consumer of psychometrics, one*

has only to show him a perfect scale and the recognition is almost instantaneous, "Yes, that's what I want."

(p. 284)

Guttman scaling provides a deterministic framework for examining dimensionality, while Rasch measurement theory can be viewed as a probabilistic extension of Guttman scaling theory. A scale is unidimensional if the properties of basic ordering of person-invariant item calibrations and item-invariant measurement are met in the data set.

One of the major implications of meeting the requirements of invariant measurement is that researchers can create a unidimensional scale. With the Rasch model, both items and persons are simultaneously located on an underlying latent scale. As shown earlier, the 2PL model implies different location orderings of items and persons with different implications for the interpretation, meaning, and use of the scores obtained on a scale.

Later in this book, we demonstrate methods for checking if the requirements of Rasch measurement theory have been achieved. Our approach to model-data fit depends mainly on the use of various types of residual analyses to evaluate the requirements of invariant measurement for a particular data set.

1.3 Components of Scale Development Based on Rasch Measurement Theory

The process by which concepts are translated into empirical indices has four steps: an initial imagery of the concept, the specification of dimensions, the selection of observable indicators, and the combination of indicators into indices.

(Lazarsfeld, 1958, p. 109)

One of the most important tasks in the social, behavioral, and health sciences is the development of substantive theories of human behavior based on a well-developed, understandable, and agreed-upon set of concepts. The "initial imagery of the concept" (Lazarsfeld, 1958, p. 109) plays a central role in our theories in the human sciences. As pointed out by Lazarsfeld (1966), "problems of concept formation, of meaning, and of measurement necessarily fuse into each other" (p. 144).

It can be argued that the measurement aspects of social science research are the least developed and understood area of research with great potential for confounding our understanding of human behavior and action. If there is not a clear understanding of units and the meaning of our measures, then it is virtually impossible to have any practical plan for implementing a theory of action based on our theoretical framework. Rasch measurement theory provides a basis for developing a set of stable measures to anchor and critically evaluate the broader theoretical models of human behavior.

Stone, Wright, and Stenner (1999) describe how measurement is made by analogy, and that maps can provide visual representations of latent variables. They highlight how useful other maps, such as rulers (length), clocks (time), and thermometers (temperature), are for communicating an underlying latent variable by analogy. In their words, "Successful item calibration and person measurement produce a map of the variable. The resulting map is no less a ruler than the ones constructed to measure length" (Stone et al., 1999, p. 321).

A Wright map represents a Rasch scale developed to measure a latent variable (e.g., food insecurity, math proficiency, and learning motivation), and it should be viewed as both a set of hypotheses regarding the latent variable and ultimately a validated definition of the latent variable defined by the scale. Wright maps have also been called variable maps and item maps. Wilson (2011) proposed calling them *Wright maps* in order to honor Professor Benjamin D. Wright at the University of Chicago who was a major figure in expanding the development and use of Rasch measurement theory around the world (Wilson & Fisher, 2017).

Figure 1.5 presents one of the earliest item maps created by Thurstone (1927) to represent severity of crimes. Several features should be noted. First, Thurstone uses a line as his map for the latent variable of seriousness of offenses. Next, he provides a scaling of the particular items on the latent variable scale. And finally, he provides a conceptual grouping of the items into sex offenses, property offenses, and injury to the person in order to highlight an underlying structure (domains and dimensions for grouping the crimes). As another example, Figure 1.6 provides a Wright map that shows the simultaneous ordering of both items and persons on the latent variable of food insecurity. These data are used throughout this book to illustrate the construction of a scale based on the principles of Rasch measurement theory.

Figure 1.7 presents the components of scale development based on Rasch measurement theory. Central to all of the components is the definition of a latent variable that is manifested in a scale that can be shared and used by a community of scholars. All of the components

Figure 1.5   Thurstone's Item Map for Seriousness of Criminal Offenses

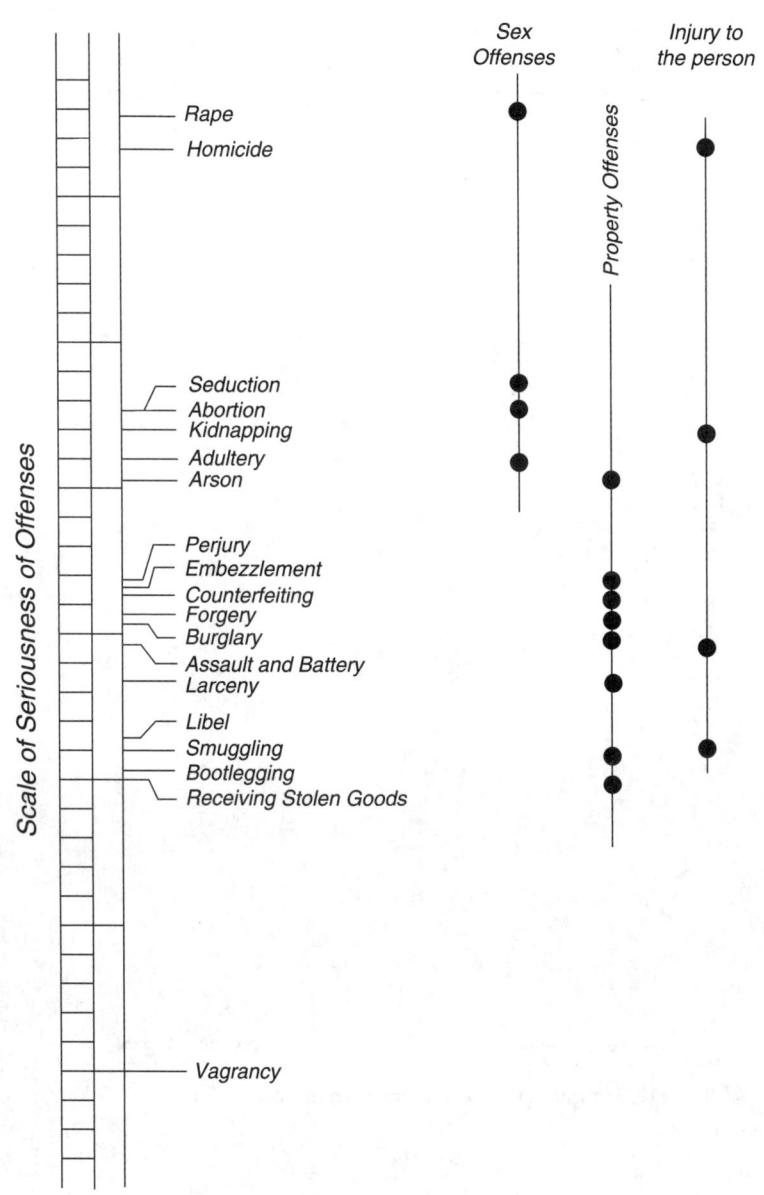

Source: Thurstone (1927).

16

Figure 1.6 Wright Map for Food Insecurity Experience Scale

```
|-------------------------------------------------|
|Logit|+Person                       |-Items      |
|----+--------------------------------+-----------|
|     | High food insecurity         | Hard to affirm      |
|  3 +                                +            |
|     |                               | Whole Day  |
|     | 22 23 29                      |            |
|     |                               |            |
|     |                               |            |
|  2 +                                +            |
|     |                               |            |
|     |                               |            |
|     | 13 17 25 33 37                |            |
|     |                               |            |
|     |                               |            |
|  1 +                                +            |
|     |                               |            |
|     |                               | Hungry     |
|     | 26 35                         |            |
|     |                               |            |
|     |                               | Ran Out  Skipped  Worried|
|  *  0 * 2 7 9 21                    *            *|
|     |                               |            |
|     |                               | Ate Less   |
|     |                               |            |
|     | 1 11 24 27 28 36              |            |
| -1 +                                +            |
|     |                               |            |
|     | 3 4 12 18 20 39               |            |
|     |                               | Healthy    |
|     |                               |            |
| -2 +                                +            |
|     |                               |            |
|     |                               | Few Foods  |
|     | 5 6 8 10 14 15 16 19 30 31 32 34 38 40 |   |
|     |                               |            |
| -3 +                                +            |
|     | Low food insecurity          | Easy to affirm    |
|-------------------------------------------------|
```

Note. The full text for the eight items is described in Chapter 2.

Figure 1.7   Components of Scale Development

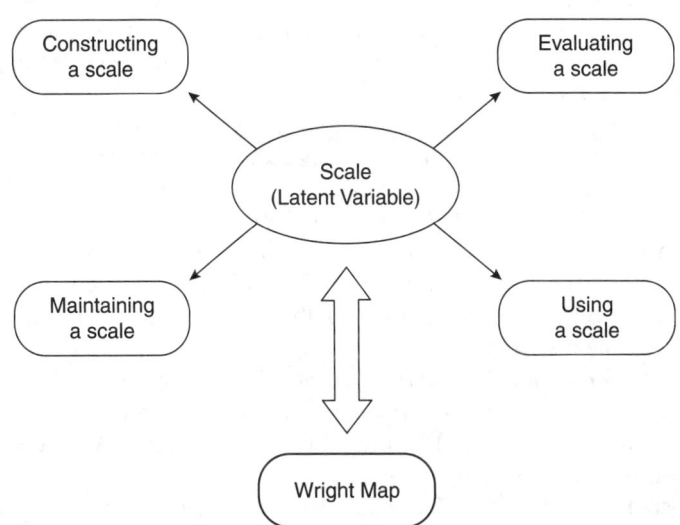

contribute to the scale that is created to represent the latent variable. Within the context of Rasch measurement theory, the Wright map plays a key role for describing, representing, and defining the latent variable. It should be noted that the components are interconnected, but that each component is likely to be stressed at different times during the scale development process.

The first component in scale development is *constructing the scale*. The guidelines used for this component are inspired by Wilson (2005). We have modified his constructing measures approach in some ways, but the key idea of a construct map (Wright map) is fundamental for this component. We adapt the idea of four building blocks (Wilson, 2005). These four building blocks are the construct map, the item design, the outcome space, and the measurement model. Each of these building blocks is described in detail later in our book with an example of constructing a Food Insecurity Experience (FIE) scale.

Once a scale is constructed, the second component involves *evaluating the scale*. Essentially, this component focuses on empirical examinations of how well the requirements of invariant measurement are met with a particular data set. Rasch measurement theory provides a set of specific requirements that offer the opportunity to achieve item-invariant measurement, person-invariant item calibration, and the creation of a unidimensional scale (Wright map). We use an approach based on the analyses of residuals to critically evaluate model-data fit (Wells & Hambleton, 2016).

The next component is *maintaining the scale*. Researchers seek to develop stable scales that meet the demands of invariant measurement, especially ensuring comparability of measures obtained in different conditions. Once a scale is created and evaluated, it is important to develop a plan and set of principles for maintaining the robustness and comparability of the scores. Some of the main topics include what has been called equating and linking of the scales based on Rasch measurement theory.

The final component is *using the scale*. The reliability, validity, and fairness of the scores obtained from a scale must be examined. Also, it is important to apply the scale in policy contexts that include the setting of performance standards.

It should be stressed that these four components are not always enacted in a strict fashion or in a specified order, and that different components have different levels of emphasis at different stages in scale development. It should also be highlighted that there are interconnections between the various components, and that the components reflect an iterative process of scale development.

1.4 Four Measurement Problems

There are a variety of measurement problems encountered in the social sciences. Many of these problems can be addressed in terms of invariant measurement with Rasch measurement theory as one approach. In this section, we introduce the use of Rasch measurement theory as an aid in conceptualizing four important measurement problems. These measurement problems are developed in greater detail in each chapter of our book. Specifically, we consider the following measurement problems:

- Definition of a latent variable
- Evaluation of differential item functioning
- Examination of interchangeability of items for person measurement
- Creation of performance standards (cut scores) for standard setting

One of the basic problems in the social sciences is the selection and definition of latent variables based on substantive theories. How do researchers select their variables? How do researchers select and calibrate various indicators to measure these variables? What guidelines should a community of scholars use to evaluate the quality of the proposed measures? We argue in this book that the use of Rasch measurement theory to create meaningful and useful scales is essential for advancing research in the social, behavioral, and health sciences. The use of arbitrary units (e.g., setting the unit based on an estimate of variance in an arbitrary sample) is unsatisfactory especially when there are clear guidelines for creating scales with meaningful units based on Rasch measurement theory.

Differential item functioning is a measurement problem that violates measurement invariance of a scale (Millsap, 2011). As pointed out by Millsap (2011), "measurement invariance is built on the notion that a measuring device should function the same way across varied conditions, as long as those varied conditions are irrelevant to the attribute being measured" (p. 1). In this book, we view this problem as a failure to meet the requirement of person-invariant calibration of items.

The next measurement problem highlights the interchangeability of items. It is important to recognize the duality of scaling models (Engelhard, 2008a). Measurement invariance is reflected in differential item functioning, but this idea is also relevant for obtaining comparable scores for persons using different subsets of items. The item-invariant person measurement provides the opportunity to achieve measure invariance for persons being independent of specific items. This measurement problem falls under the general category of test equating—the main idea is that the latent variable and the Wright map that represents the scale remain invariant over different sets of observations defined by different items but designed to define the same construct for persons. Computer-adaptive testing is based on the idea that different items are interchangeable, and different subsets of items can be used to obtain comparable measures for each person.

The final measurement problem that we address is standard setting (Cizek, 2012). Standard setting is a process for determining critical points on a scale (cut scores) that represent substantively distinct locations along the latent variable. In many cases, such as educational achievement and food insecurity, ordered categorical reporting of continuous scales is required (fail/pass, food secure/food insecure) to inform policy. Standard setting is the process used to set these levels and create meaningful categories to guide policy makers and other stakeholders.

The chapters of this book provide illustrations of how Rasch measurement theory can be used to solve measurement problems through the components of scale development. Our aim is to describe an approach for problem-solving in measurement emerging from the requirements for invariant measurement and the quest for robust and stable systems of measurement.

2

CONSTRUCTING A RASCH SCALE

Rasch measurement theory can be used as the basis for solving a variety of measurement problems. In the previous chapter, we proposed using four components as the basis for developing a Rasch scale: constructing, evaluating, using, and maintaining a scale. Each of these components maps to a common measurement problem encountered in the social sciences. These problems are the definition of a latent variable, measurement invariance (e.g., differential item functioning), interchangeability of items (e.g., test equating), and standard setting (e.g., setting a cut score based on performance standards). This chapter discusses the first component in relation to the measurement problem of constructing a scale that can be used to define a latent variable based on the principles of Rasch measurement theory.

Once a researcher has decided to use a scale to represent an important latent variable (construct), the first step is to begin the construction of the scale. The construction of a scale involves several steps. The approach used here is based on modifications to the constructing measures approach suggested by Wilson (2005). The essential building blocks for constructing a latent variable scale include specification of the latent variable (construct) to be measured, creation of an observational design (e.g., items or questions), development of a set of scoring rules, and application of the Rasch model to observed data to create an empirical Wright map.

This chapter begins with a description of the building blocks that can be used to create a Rasch scale. Next, we use an international scale constructed to measure individual experiences of food insecurity as an illustrative example (Cafiero, Viviani, & Nord, 2018). This includes the use of the Rasch model to evaluate a small data set. Finally, we provide a summary of the chapter and highlight key points.

2.1 Building Blocks for a Rasch Scale

The creation of a meaningful and useful scale based on Rasch measurement theory is described using the four building blocks as shown in Figure 2.1 (latent variable, observational design, scoring rules, and

Figure 2.1 Building Blocks for Constructing a Rasch Scale

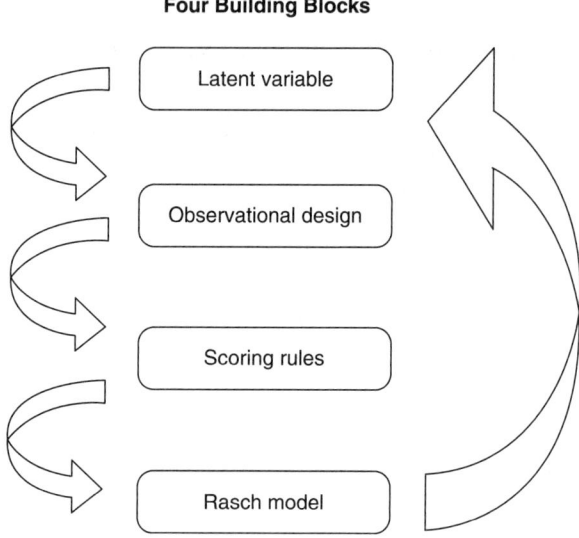

Four Building Blocks

Latent variable

Observational design

Scoring rules

Rasch model

Rasch model). The specific question addressed is: What are the essential steps for constructing a Wright map based on Rasch measurement theory? Each building block is described in detail in this section.

Latent Variable

The first building block in creating a scale starts with the initial imagery of a latent variable (Lazarsfeld, 1958). It is important to note that we are creating a unidimensional scale. Unidimensional scales play key roles because "they coincide with the use of unidimensional language in social science theories—language that is intended to clarify the meaning of those theories" (McIver & Carmines, 1981, p. 86). The concept of unidimensionality is relative in essence, e.g., if the items measure both mathematical and reading components to the same degree, the items may be scalable on a unidimensional scale (Andrich, 1985; Lumsden, 1957).

The measurement of food insecurity is used for our illustrations in this chapter. The purpose of the Food Insecurity Experience (FIE) scale is to obtain evidence regarding food insecurity in a global context

(Cafiero et al., 2018). Overall, food insecurity is defined very generally as follows:

> *Food security is said to exist when all people, at all times, have physical, social and economic access to sufficient safe and nutritious food that meets their dietary needs and food preferences for an active and healthy life.*
>
> Declaration of the World Summit on Food Security, World Summit on Food Security, Rome, November 16–18, 2009

Cafiero et al. (2018) recognize that:

> *although food security is inherently multi-dimensional, one critical dimension is continued access to adequate food. The United Nations Food and Agriculture Organization (FAO) has undertaken a project called Voices of the Hungry (VoH) to develop and support a survey-based experiential measure of access to food, called the Food Insecurity Experience Scale.*
>
> (p. 146)

A similar approach is used in the United States for measuring food insecurity at the household level (Coleman-Jensen, Rabbitt, Gregory, & Singh, 2015). The FIE scale measures food insecurity conceived as the "condition of not being able to freely access the food one needs to conduct a healthy, active and dignified life ... resulting from the inability to access food due to lack of money or other resources" (p. 147). Narrowing down the broad theoretical definition of food insecurity to focus on a single dimension, access to adequate food guides the creation of a hypothesized Wright map for measuring food insecurity.

The hypothesized Wright map for food insecurity is shown in Figure 2.2. There are several features that should be noted in Figure 2.2. First, the scale for measuring food insecurity is represented by a line. This line reflects a theoretical continuum that ranges from low food insecurity to high food insecurity. This continuum includes qualitative descriptions of ordered levels of food insecurity for persons that range from mild food insecurity through moderate food insecurity to severe food insecurity. These levels represent substantively important distinctions that are used by policy makers who address issues of food insecurity around the world. Second, the line representing the latent variable of food insecurity is defined by an ordered set of items that reflect the experiences of food insecurity. These ordered items reflect a

Figure 2.2 Hypothesized Wright Map for Measuring Food Insecurity

Food Insecurity Levels		Food Insecurity Experiences (Items)
High food insecurity	▲	Hard to affirm
		Experiencing hunger
Severe food insecurity		
		Reducing quantities, skipping meals
Moderate food insecurity		
		Compromising quality and variety of food
Mild food insecurity		
		Worrying about ability to obtain food
Low food insecurity	▼	Easy to affirm

Source: Based on Coleman-Jensen et al. (2015).

hypothesized expectation regarding the order of items from easy to affirm with a Yes response (i.e., worrying about ability to obtain food) to hard to affirm (i.e., experiencing hunger).

Observational Design

After a researcher has created a hypothesized Wright map (e.g., Figure 2.2), the next step is the creation of a set of observable indicators or items to represent food insecurity. Observational designs frequently include various item classifications and domains that guide the creation of specific items. A classic example is the creation of educational achievement tests using item classifications based on Bloom's Taxonomy (Bloom, Engelhart, Furst, Hill, & Krathwohl, 1956). Lane, Raymond, and Haladyna (2016) provide a detailed consideration of various guidelines for test and item development for assessments used in a variety of assessment contexts.

Table 2.1 shows the eight items that are included in the FIE scale (Cafiero et al., 2018) used in this book. These items reflect the observational design used to represent food insecurity. This scale is based on a careful consideration of previous scales that have been used to measure household and individual food insecurity around the world, such as the US Household Food Security Survey Module (HFSSM), the Escala Brasileira de Insegurança Alimentar (EBIA), the Escala

Table 2.1 English Version of the Food Insecurity Experience Scale

Item	Questions	Label
1	During the last 12 months, was there a time when you were worried you would not have enough food to eat because of a lack of money or other resources?	Worried
2	Still thinking about the last 12 months, was there a time when you were unable to eat healthy and nutritious food because of a lack of money or other resources?	Healthy
3	Was there a time when you ate only a few kinds of foods because of a lack of money or other resources?	Few Foods
4	Was there a time when you had to skip a meal because there was not enough money or other resources to get food?	Skipped
5	Still thinking about the last 12 months, was there a time when you ate less than you thought you should because of a lack of money or other resources?	Ate Less
6	Was there a time when your household ran out of food because of a lack of money or other resources?	Ran Out
7	Was there a time when you were hungry but did not eat because there was not enough money or other resources for food?	Hungry
8	During the last 12 months, was there a time when you went without eating for a whole day because of a lack of money or other resources?	Whole Day

Note. Respondents answer yes or no to these questions (http://www.fao.org/in-action/voices-of-the-hungry/fies/en/).

Latinoamericana y Caribeña de Seguridad Alimentaria (ELCSA), the Escala Mexicana de Seguridad Alimentaria (EMSA), and the Household Food Insecurity Access Scale (HFIAS) (Coleman-Jensen et al., 2015). The selection of items also included a consideration of the interpretability of these items and conditions across different cultures and contexts by the creators of the scale.

Scoring Rules

Scoring rules specify how the person responses are coded. For our example, the responses to the eight items are simply scored dichotomously (Yes = 1 and No = 0). A response of Yes indicates an affirmative response to the item, and it leads to a higher level of food insecurity. The items are combined into sum scores with higher sum scores indicating more severe food insecurity of persons.

There are other examples of scoring rules that include different types of rating scales, such as the rating scale model (Andrich, 2016) and partial credit model (Masters, 2016) that are part of the Rasch family of models (Wright & Masters, 1984). It is also possible to combine categories in ways that reflect different scoring rules. Engelhard and Wind (2018) provide guidance on different models for polytomous or rating data using different Rasch models.

Rasch Model

The final step is the use of a measurement model to link the observed responses to items and persons based on their locations on a latent variable scale. Rasch (1960/1980) started with a simple idea that a person's response to an item depends on the difficulty of the item and the ability of the person. He selected a probabilistic model based on the logistic response function because of its desirable properties related to specific objectivity (i.e., invariant measurement). The dichotomous Rasch model in its modern form can be written as:

$$\phi_{ni1} = \frac{\exp(\theta_n - \delta_{i1})}{1 + \exp(\theta_n - \delta_{i1})} \tag{2.1}$$

The Rasch model in Equation 2.1 can be viewed as an operating characteristic function that relates the differences between locations of persons (θ) and items (δ) on a latent variable to the probability of success or affirmation on a dichotomous item. This distance reflects a comparison between each person and item that predicts a probability of a positive response for each person on an item.

We find it useful to conceptualize the Rasch model as a probabilistic version of Guttman scaling (Andrich, 1985). In his words:

... technical parallels between the SLM [Rasch model] and the Guttman scale are not a coincidence. The connections arise from the same essential conditions required in both, including the requirement of invariance of scale and location values with respect to each other.

(Andrich, 1988, p. 40)

Engelhard (2005) describes Guttman scales in detail as deterministic- and ideal-type models. Table 2.2 provides a simple example to show the

Table 2.2 Illustration of Guttman (Perfect) and Rasch (Probabilistic) Item Response Patterns

	Panel A				Panel B			
	Perfect Pattern (Guttman)				Probabilistic Pattern (Rasch)			
Person Scores	A Easy	B	C	D Hard	A Easy	B	C	D Hard
4	1	1	1	1	0.98	0.95	0.75	0.65
3	1	1	1	0	0.95	0.75	0.65	0.45
2	1	1	0	0	0.75	0.65	0.45	0.34
1	1	0	0	0	0.65	0.45	0.34	0.25
0	0	0	0	0	0.45	0.34	0.25	0.15

Note. These values are used for illustration.

triangular pattern of Guttman scales. Panel A in Table 2.2 shows this pattern when items are ordered from easy to hard and the persons are ordered based on their sum scores. A similar triangular pattern appears with the Rasch probabilities when items are calibrated, and persons measured on the latent variable scale. This is illustrated in Panel B of Table 2.2. Probabilities that are greater than 0.50 indicate that a person is more likely to affirm an item, although the stochastic nature of the process also recognizes that a person may not affirm the item. This is in contrast to Guttman scaling that defines a perfect scale with a deterministic model.

Figure 2.3 provides a flowchart of the progress from the idea of a latent variable (construct) to a Wright map. Once we have a general conception of the latent variable, the next steps are to create an observational design and scoring rules that guide the creation of items and indicators that we plan to use to define the latent variable. Next, the items are administered to a sample of persons to collect responses, and the observed data are analyzed with the Rasch model. The Rasch model is the measurement model that connects the observed data to the visual representation of our latent variable on the Wright map. Finally, the Wright map displays the location of both items and persons empirically on the latent variable scale.

There are numerous excellent introductions to the technical details for estimating item and person locations on the latent variable based on the Rasch model. We highly recommend Baker and Kim (2004) as an advanced text addressing methods for estimating the parameters of item

Figure 2.3 Progression From Latent Variable (Construct)
to Wright Map

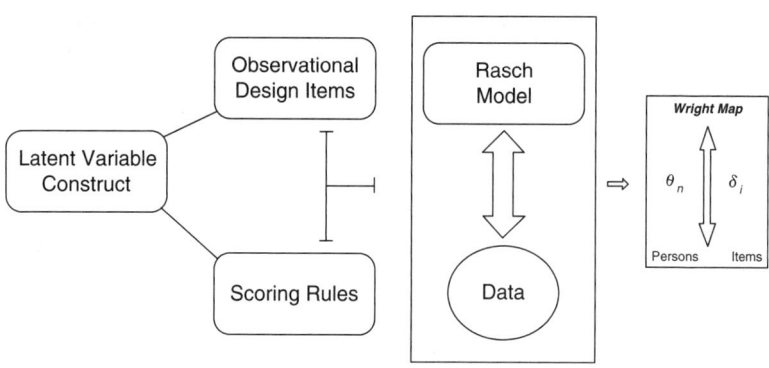

response theory (IRT) models. Baker and Kim (2017) have also published R syntax for a variety of IRT models.

In this book, we use the Facets computer program (Linacre, 2018a) to estimate the parameters of the Rasch model. The syntax for the Facets computer program and also an R program (Everyone's Rasch Measurement Analyzer—ERMA) is available online (https://study.sagepub.com/researchmethods/qass/engelhard-rasch-models). Sample data sets are also available online.

2.2 Illustrative Analyses

In this section, we use a dichotomous Rasch model to analyze the FIE data. The dichotomous responses of 40 persons to 8 items (FIE scale) are shown in Table 2.3. These data reflect food insecurity experiences for the United States. Table 2.4 presents the summary statistics for the eight items from a Rasch analysis of these data responses using the Facets computer program (Linacre, 2018a). The first column is the item number, while the second column provides a short label describing the content of the item.

Column 3 shows the proportion of Yes responses to each item. The items range from easy to affirm (Item 3—Few Foods) to hard to affirm (Item 8—Whole Day). The next two columns indicate the calibration of the items in logits (and standard errors) that represent the locations of the items on the Wright map. The Wright map is shown in Chapter 1 (Figure 1.6). The next four columns report how well the observed data

Table 2.3 Food Insecurity Experience Data

Person	Items							
	1	2	3	4	5	6	7	8
1	1	0	1	0	1	0	0	0
2	0	0	1	1	1	0	1	0
3	0	0	1	0	1	0	0	0
4	0	0	0	0	0	1	1	0
5	0	0	1	0	0	0	0	0
6	0	0	1	0	0	0	0	0
7	1	1	1	0	1	0	0	0
8	0	0	1	0	0	0	0	0
9	1	1	1	0	0	1	0	0
10	0	0	1	0	0	0	0	0
11	0	1	1	0	0	1	0	0
12	0	1	0	0	0	1	0	0
13	1	1	1	1	1	1	0	0
14	0	1	0	0	0	0	0	0
15	0	1	0	0	0	0	0	0
16	0	0	1	0	0	0	0	0
17	0	1	1	1	1	1	1	0
18	0	1	1	0	0	0	0	0
19	0	0	1	0	0	0	0	0
20	0	1	1	0	0	0	0	0
21	0	1	1	1	0	1	0	0
22	1	1	1	1	1	1	1	0
23	1	1	0	1	1	1	1	1
24	0	1	1	0	1	0	0	0
25	1	1	1	1	0	0	1	1
26	1	1	1	1	0	1	0	0
27	1	0	1	0	1	0	0	0
28	0	0	1	1	0	0	1	0
29	0	1	1	1	1	1	1	1
30	0	0	0	0	1	0	0	0

Table 2.3 *(Continued)*

| | | | | | Items | | | |
Person	1	2	3	4	5	6	7	8
31	0	1	0	0	0	0	0	0
32	0	1	0	0	0	0	0	0
33	1	1	1	0	1	1	1	0
34	1	0	0	0	0	0	0	0
35	1	1	1	1	1	0	0	0
36	0	0	0	1	1	0	1	0
37	1	1	1	1	1	1	0	0
38	0	1	0	0	0	0	0	0
39	0	1	1	0	0	0	0	0
40	0	0	1	0	0	0	0	0

Note. Items are scored as follows: 0 = No, 1 = Yes.
Source: Based on Coleman-Jensen et al. (2015).

Table 2.4 Summary Statistics for Items (Ordered by Proportion of Yes Responses)

					Mean Squares		Fit Category	
Item	Label	Proportion of Yes Responses	Measure	S.E.	Infit	Outfit	Infit	Outfit
8	Whole Day	0.08	2.85	0.67	0.74	0.23	A	B
7	Hungry	0.25	0.81	0.46	1.04	0.81	A	A
1	Worried	0.33	0.21	0.43	1.02	1.16	A	A
4	Skipped	0.33	0.21	0.43	0.69	0.47	A	B
6	Ran Out	0.33	0.21	0.43	0.92	0.71	A	A
5	Ate Less	0.40	−0.32	0.41	1.03	0.93	A	A
2	Healthy	0.60	−1.59	0.39	1.20	0.98	A	A
3	Few Foods	0.73	−2.38	0.41	1.12	4.42	A	D

Note. Fit categories: A ($0.50 \leq MSE < 1.50$), B ($MSE < 0.50$), C ($1.50 \leq MSE < 2.00$), D ($MSE \geq 2.00$).
MSE, mean square error.

Table 2.5 Fit Category Based on Mean Square Error (*MSE*)

MSE	Interpretation	Fit Category
$0.50 \leq MSE < 1.50$	Productive for measurement	A
$MSE < 0.50$	Less productive for measurement, but not distorting of measures	B
$1.50 \leq MSE < 2.00$	Unproductive for measurement, but not distorting of measures	C
$2.00 \leq MSE$	Unproductive for measurement, and distorting of measures	D

fit the Rasch model. The Infit statistics are sensitive to unexpected responses to items that are close to person locations, while the Outfit statistics are sensitive to responses to items that are located farther from person locations. Table 2.5 shows a framework suggested by Engelhard and Wind (2018) to further categorize the items.

Using this framework, none of the items misfit based on the Infit statistics, while Item 3 (Healthy) fits based on the Infit statistic and misfits based on the Outfit statistics. Further details on the Infit and Outfit statistics will be discussed in more detail in Chapter 3.

Table 2.6 provides similar information for persons. These analyses provide a validation of the responses of each person. Food insecurity ranges from high for Person 22 who responds "yes" to 88% of the items to low for Person 34 who responds "yes" to 13% of the items. As with the items, the measures indicate the location of the persons in logits on the Wright map (Figure 1.6).

Based on the Infit statistics, a summary of the fit categories for the persons are as follows: A (75.0%), B (10.0%), C (7.5%), and D (7.5%). The fit categories based on the Outfit statistics are as follows: A (37.5%), B (47.5%), C (5.0%), and D (10.0%). Misfitting persons will be discussed in more detail in Chapter 3.

Rasch measurement theory provides the connection between the observed data and the creation of a scale including location parameters for items and persons. A Wright map provides the visual outcome of this process. There are two representations for a Wright map with the hypothesized map for food insecurity (Figure 2.2) and the empirical map shown in Chapter 1 (Figure 1.6).

Table 2.6 Summary Statistics for Persons (Ordered by Proportion of Yes Responses)

				Mean Squares		Fit Category	
Person	Prop Yes	Measure	S.E.	Infit	Outfit	Infit	Outfit
22	0.88	2.63	1.23	0.36	0.16	B	B
29	0.88	2.63	1.23	1.80	1.61	C	C
23	0.88	2.63	1.23	2.02	9.00	D	D
13	0.75	1.47	0.96	0.59	0.41	A	B
37	0.75	1.47	0.96	0.59	0.41	A	B
17	0.75	1.47	0.96	0.81	0.64	A	A
33	0.75	1.47	0.96	0.81	0.64	A	A
25	0.75	1.47	0.96	2.02	1.83	D	C
35	0.63	0.67	0.85	0.72	0.54	A	A
26	0.63	0.67	0.85	0.88	0.72	A	A
7	0.50	−0.02	0.83	0.69	0.55	A	A
9	0.50	−0.02	0.83	0.87	0.68	A	A
21	0.50	−0.02	0.83	0.87	0.68	A	A
2	0.50	−0.02	0.83	1.40	1.35	A	A
24	0.38	−0.72	0.86	0.54	0.44	A	B
11	0.38	−0.72	0.86	0.72	0.61	A	A
1	0.38	−0.72	0.86	1.16	0.95	A	A
27	0.38	−0.72	0.86	1.16	0.95	A	A
28	0.38	−0.72	0.86	1.49	1.40	A	A
36	0.38	−0.72	0.86	2.13	2.13	D	D
39	0.25	−1.51	0.95	0.40	0.29	B	B
18	0.25	−1.51	0.95	0.40	0.29	B	B
20	0.25	−1.51	0.95	0.40	0.29	B	B
3	0.25	−1.51	0.95	0.92	0.68	A	A
12	0.25	−1.51	0.95	1.39	1.21	A	A
4	0.25	−1.51	0.95	2.16	2.49	D	D
6	0.13	−2.60	1.18	0.55	0.24	A	B
8	0.13	−2.60	1.18	0.55	0.24	A	B
10	0.13	−2.60	1.18	0.55	0.24	A	B
5	0.13	−2.60	1.18	0.55	0.24	A	B

Table 2.6 *(Continued)*

| Person | Prop Yes | Measure | S.E. | Mean Squares | | Fit Category | |
				Infit	Outfit	Infit	Outfit
16	0.13	−2.60	1.18	0.55	0.24	A	B
19	0.13	−2.60	1.18	0.55	0.24	A	B
40	0.13	−2.60	1.18	0.55	0.24	A	B
14	0.13	−2.60	1.18	1.04	0.48	A	B
15	0.13	−2.60	1.18	1.04	0.48	A	B
31	0.13	−2.60	1.18	1.04	0.48	A	B
32	0.13	−2.60	1.18	1.04	0.48	A	B
38	0.13	−2.60	1.18	1.04	0.48	A	B
30	0.13	−2.60	1.18	1.52	1.39	C	A
34	0.13	−2.60	1.18	1.62	2.25	C	D

Note. Fit categories: A ($0.50 \leq MSE < 1.50$), B ($MSE < 0.50$), C ($1.50 \leq MSE < 2.00$), D ($MSE \geq 2.00$).
MSE, mean square error.

2.3 Summary

This chapter introduces the essential steps of scale construction based on Rasch measurement theory. The construction of a Rasch scale can be reflected by a flowchart that is shown in Figure 2.3. The first step includes the conceptual formation of a latent variable that is a focus of theory and practice within a broader substantive theory. We use the example of food insecurity as our focal latent variable. The next step guides our selection of items and observations for the design of the scale that is used to operationally define the latent variable. As an illustrative example, eight items (Table 2.1) are used to define the FIE scale. Meanwhile, a set of scoring rules is developed to code the observations onto an ordinal scale. The responses to the FIE scale are scored dichotomously (0 = no, 1 = yes). The last step links the Rasch model to observed data that are collected based on the responses of persons to the items. The Rasch model connects the observed data to the cali-bration of the items (location of items on the scale) and the measure-ment of persons (location of persons on the scale). The outcome of this step includes the creation of a Wright map that shows the simultaneous location of persons and items on the Rasch scale.

The empirical Wright map for the illustrative data is shown in Figure 1.6. It is important to remember that our goal is to ultimately create a scale that is validated for its intended purposes and uses. It is also important that the research community accepts the scale as a consensus view of the construct being measured by the scale. We think of this process as being organized around a Wright map with two aspects: a hypothesized Wright map and an empirical Wright map. The FIE scale provides a good example of a scale that has been widely recognized, and it is used throughout the world to measure food insecurity (Cafiero et al., 2018).

A Rasch scale meets the requirements of invariant measurement when good model-data fit is obtained. One point of confusion in the literature on psychometrics is the failure to adequately distinguish between the unobservable latent variable that is hypothesized a priori (hypothesized Wright map) and the empirical analyses of model-data fit to determine whether or not our intentions are realized in our specific data set. We view the examination of model-data fit as part of the evaluation of whether or not a successful scale for our latent variable (empirical Wright map) has been constructed. Chapter 3 describes in more detail the steps for evaluating model-data fit for a Rasch scale including item and person fit analyses.

3

EVALUATING A RASCH SCALE

Evaluating the psychometric quality of a scale after it is constructed is essential. This chapter shows several approaches for evaluating a Rasch scale to address a key question—*how well does empirical data meet the requirements of Rasch measurement theory*? The invariant measurement properties of Rasch measurement theory can be achieved only when there is good model-data fit. Model-data fit information within the context of Rasch measurement theory includes fit indices of individual items and persons, as well as person-invariant item calibration and item-invariant person measurement on a unidimensional scale. This chapter presents commonly used fit indices and procedures for detecting misfit item and person. The appropriate use of a Rasch scale and the interpretation of scores greatly depend on the achievement of invariant measurement.

This chapter provides an overview of model-data fit and the concept of invariant measurement. First, individual fit statistics for diagnosing misfit of persons and items are suggested. The analyses of residuals are presented as an approach to examine item fit and person fit in detail. Next, the invariant calibration of items across subgroups of persons is discussed through an examination of differential item functioning. Differential item functioning is a potential measurement problem that may distort the invariant measures on the scale, so that must be examined empirically. The next section introduces the requirement of unidimensionality and its importance for achieving invariant measurement. An illustrative analysis is used to evaluate model-data fit for the Food Insecurity Experience (FIE) scale through residual analysis and differential item functioning. Finally, we summarize the major points discussed in this chapter.

3.1 Rasch's Specific Objectivity

In the early 1960s, Rasch (1960/1980) introduced an epistemological concept and named it *objectivity*. Later on, he added a restricting predicate—*specific* for objectivity being used particularly to evaluate a measurement matter. Rasch (1960/1980) described specific objectivity as a unique property of Rasch measurement theory that seeks for sample-free item calibration and test-free person measurement (Wright, 1968). In particular, the specific objectivity can be defined as follows. On one hand, the comparison between items should be invariant across particular individuals who were sampled for the comparison. On the other hand, the comparison between persons should also be invariant across subsets of items that were instrumental for the comparison. Rasch (1977) suggests that the comparisons between items or persons under such circumstances or that fulfill these requirements are *specifically objective*. Engelhard (2013) further discusses the concept of specific objectivity within a framework of invariant measurement and emphasizes the importance of creating a unidimensional scale for achieving item invariance across subpopulation groups and person invariance across subsets of items. A unidimensional scale refers to a single continuum that can simultaneously show relative locations of persons and items, and this continuum can be empirically visualized through a Wright map.

Based on Rasch measurement theory, person-invariant item calibration is evidenced by noncrossing item response functions. Noncrossing item response functions ensure that a person always has a higher probability of endorsing an easier item than a harder one. Meanwhile, noncrossing person response functions reflect the requirement of item-invariant person measurement. Noncrossing person response functions show that the probability of endorsing an item is always higher for a person with higher measure on the latent variable than a person with lower measure. This is consistent with the basic notion of Guttman scaling. As described in Chapter 1, a Rasch scale can be viewed as a probabilistic Guttman scale.

We illustrated the construction of a FIE scale based on Rasch measurement theory in Chapter 2. In this particular case, any individual would have higher probability of affirming Item 3 (Was there a time when you ate only a few kinds of food because of a lack of money or other resources?) than Item 8 (During the last 12 months, was there a time when you went without eating for a whole day because of a lack of money or other resources?), as Item 3 reflects less severe food insecurity condition. The rank order of the items based on the relative difficulty to affirm is invariant for any individual. On the other hand, Person 33 who has the highest food insecurity measure would have higher probability of endorsing an item than Person 40 who has the lowest measure of food insecurity. The noncrossing curves again define the invariant measurement properties of Rasch measurement theory that are crucial to any objective scale.

3.2 Model-Data Fit

The evaluation of a Rasch scale essentially involves examining the degree to which invariant measurement has been achieved based on empirical data. Rasch (1960/1980) suggested thinking about whether the model or data has gone wrong when discordance appears related to model-data fit. For scale construction purposes, items should function like a *tick mark* on a yardstick (i.e., scale for latent variable of length).

Assessing the fit between data and a Rasch scale is analogous to examining the amount of deviation from the perfect reproducibility of an ideal Guttman pattern (Engelhard, 2005). In this case, the residual-based fit indices can be used to quantify and interpret the deviations between model and data. Early work on evaluating model-data fit began with an error-counting method proposed by Edwards (1948). This approach compares the observed responses to an ideal Guttman response pattern and simply counts the number of different errors in the empirical data. Edwards (1948) developed this error-counting method based on Guttman's (1947, 1950) definition of perfect reproducibility (Engelhard, 2005; McIver & Carmines, 1981). Nowadays, Pearson's chi-square test and a likelihood ratio test can be used to assess the fit of a Rasch scale (Millsap, 2011). The individual-based fit indices for Rasch models for each item and person are also available. The development of fit indices for Rasch models is generally based on an analysis of residuals (Wells & Hambleton, 2016).

Residuals represent the distances between observed responses and model-based probabilities. Smaller residuals are preferred as they indicate better agreement between model and data. The residuals can be analyzed directly to detect misfitting items and persons. The raw residual (R_{ni}) is defined as the observed response to item i given by person n (X_{ni}) minus the model-based probability for person n to answer item i correctly (P_{ni}). Figure 3.1 shows the creation of a residual matrix of raw residuals.

The standardized residual (SR_{ni}) simply applies standardization to the raw residual dividing the raw residual by its standard error. The equations for calculating raw residuals and standardized residuals are as below:

$$R_{ni} = X_{ni} - P_{ni} \tag{3.1}$$

38

Figure 3.1 Residual Matrix

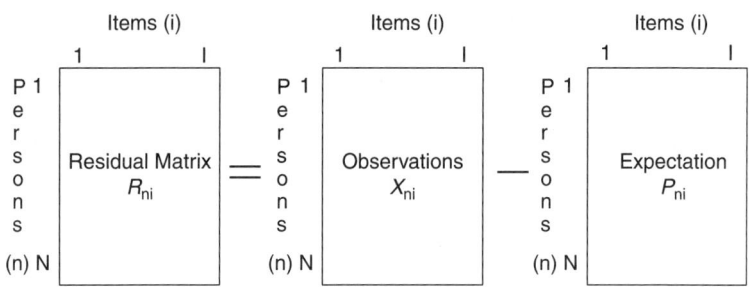

Note. Please see https://www.rasch.org/rmt/rmt231j.htm and Engelhard (2013, p. 18) for more details.

$$SR_{ni} = \frac{R_{ni}}{SE_R} = \frac{X_{ni} - P_{ni}}{\sqrt{\frac{P_{ni}(1 - P_{ni})}{N}}} \qquad (3.2)$$

where

$i = 1,...,I$ and denotes individual item,

$n = 1,...,N$ and denotes individual person,

R_{ni} = raw residual on item i by person n,

X_{ni} = observed response to item i by person n,

P_{ni} = expected probability for person n to answer item i correctly,

SE_R = standard error of raw residuals, and

SR_{ni} = standardized residual on item i by person n.

By dividing persons into equal ability intervals, we can obtain a raw residual for item i in an interval based on the observed proportions minus the model-based probability of correct response, where the model-based probability of correct responses can be either mean or median of expected probabilities of all persons in this interval who have answered item i correctly (Wells & Hambleton, 2016). The standardized residuals can be used to create residual plots to further visualize irregular patterns or outliers.

Besides the direct use of residual statistics, the Infit and Outfit statistics are widely applied for evaluating Rasch scales. The Infit and Outfit statistics further quantify the deviations between observed responses and expected probabilities. In particular, the Infit statistics are more sensitive to irregular patterns where persons/items are well targeted, whereas the Outfit statistics can better detect outliers throughout the whole range of the scale (Engelhard, 2013). The following equations show the calculation of Infit and Outfit statistics for evaluating *item fit*.

$$\text{Item Infit mean square(Infit MnSq)} : v_i = \frac{\sum\limits_{n=1}^{N} (X_{ni} - P_{ni})^2}{\sum\limits_{n=1}^{N} P_{ni}(1 - P_{ni})} \quad (3.3)$$

$$\text{Item Outfit mean square(Outfit MnSq)} : u_i = \frac{1}{N} \sum\limits_{n=1}^{N} \frac{(X_{ni} - P_{ni})^2}{P_{ni}(1 - P_{ni})} \quad (3.4)$$

$$\text{Item Infit standardized } Z : t_i = (v_i^{1/3} - 1)(\frac{3}{\sigma_v^2}) + (\frac{\sigma_v^2}{3}),$$

and

$$\text{Variance of Infit mean square} = \sigma_v^2 = \frac{\sum\limits_{n=1}^{N} (C_{ni} - \sigma_{ni}^4)}{(\sum\limits_{n=1}^{N} \sigma_{ni}^2)^2} \quad (3.5)$$

$$\text{Item Outfit standardized } Z : t_i = (u_i^{1/3} - 1)(\frac{3}{\sigma_u^2}) + (\frac{\sigma_u^2}{3}),$$

and (3.6)

$$\text{Variance of Outfit mean square} = \sigma_u^2 = \sum\limits_{n=1}^{N} (\frac{C_{ni}/\sigma_{ni}^4}{N^2}) - \frac{1}{N}$$

where σ_{ni}^2 = variance of X_{ni} and C_{ni} = kurtosis of X_{ni}.

The equations can be easily modified to calculate *person fit* indices:

$$\text{Person Infit mean square (Infit MnSq)} : v_n = \frac{\sum_{i=1}^{I} (X_{ni} - P_{ni})^2}{\sum_{i=1}^{I} P_{ni}(1 - P_{ni})}$$

(3.7)

$$\text{Person Outfit mean square (Outfit MnSq)} : u_n = \frac{1}{I} \sum_{i=1}^{I} \frac{(X_{ni} - P_{ni})^2}{P_{ni}(1 - P_{ni})}$$

(3.8)

$$\text{Person Infit standardized } Z : t_n = (v_n^{1/3} - 1)(\frac{3}{\sigma_v^2}) + (\frac{\sigma_v^2}{3}),$$

and

(3.9)

$$\text{Variance of Infit mean square} = \sigma_v^2 = \frac{\sum_{i=1}^{I} (C_{ni} - \sigma_{ni}^4)}{(\sum_{i-1}^{I} \sigma_{ni}^2)^2}$$

$$\text{Person Outfit standardized } Z : t_n = (u_n^{1/3} - 1)(\frac{3}{\sigma_u^2}) + (\frac{\sigma_u^2}{3}),$$

and

(3.10)

$$\text{Variance of Outfit mean square} = \sigma_u^2 = \sum_{i=1}^{I} (\frac{C_{ni}/\sigma_{ni}^4}{I^2}) - \frac{1}{I}$$

Figure 3.2 uses matrices to illustrate the calculation of Infit and Outfit mean square residuals. The Outfit mean square is unweighted, and it is hypothesized to follow a chi-square distribution (Linacre, 2018b). The Infit mean square is further weighted by the item information function. Both Infit and Outfit mean square statistics have an expectation of one—a closer-to-one value implies good item or person fit. The standardized Infit and Outfit statistics are hypothesized to follow a standard normal distribution with an expectation of zero and a standard deviation of one where a closer-to-zero statistic indicates good

Figure 3.2 Infit and Outfit Mean Square Residuals

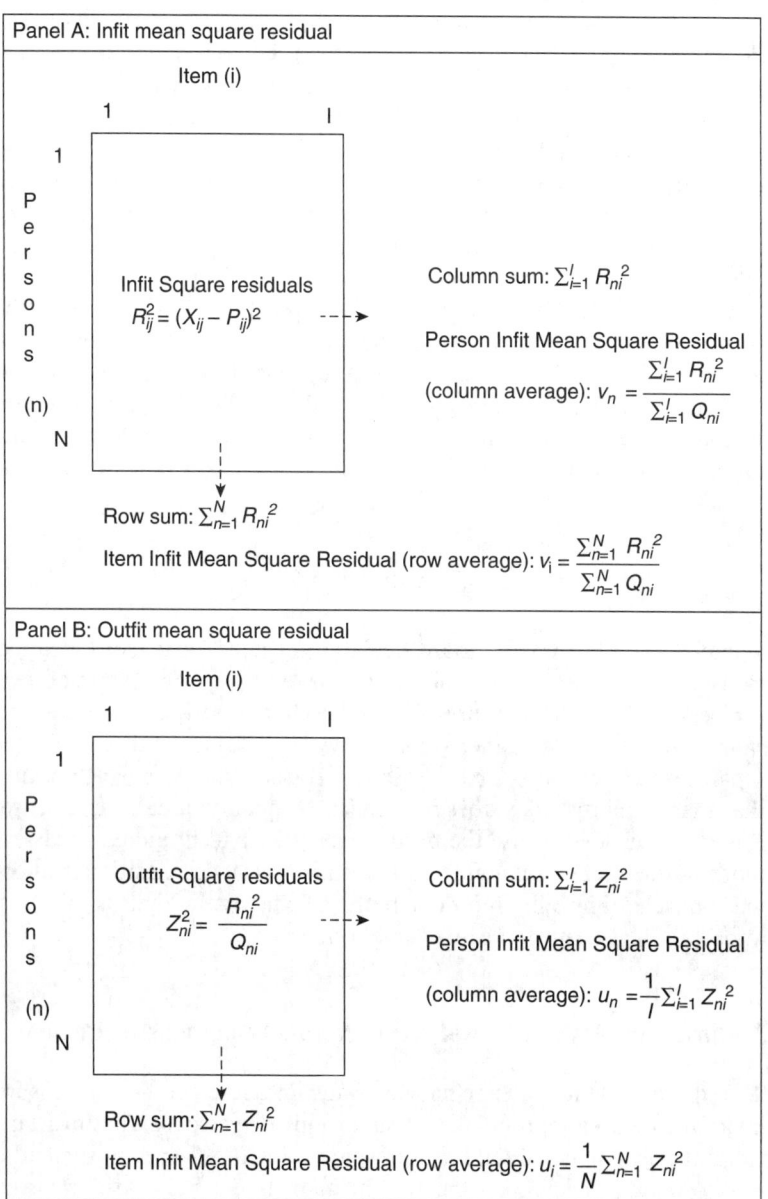

Note. Residual $R_{ni} = X_{ni} - P_{ni}$; information $Q_{ni} = P_{ni}(1 - P_{ni})$; please see https://www.rasch.org/rmt/rmt231j.htm and Engelhard (2013, p. 18) for more details.

fit. It is easier to apply standardized statistics for diagnosing the degrees of misfit since the standard deviation is one. However, standardized Infit and Outfit statistics tend to vary more with different sample sizes (Karabatsos, 2000; Smith & Hedges, 1982). In Section 3.4, we will use the illustrative example to further interpret the fit indices.

Within the context of Rasch measurement theory, the reliability of separation index reflects the reproducibility of the relative location measures. In another word, it assesses to what degree the scale can always produce the estimated location measures. We can obtain the reliability of separation index for item and person facet separately. The reliability of item separation indicates how sufficiently the items are separated in difficulty to represent the direction and meaning of the construct. Wright and Masters (1982) emphasized that "our success in defining a line of increasing intensity depends on the extent to which items are separated" (p. 91). On the other hand, the reliability of person separation reflects to what degree the instrument can separate the sample to different levels. The reliability index can be calculated as below:

$$\text{Reliability of Separation} = \frac{SD^2 - MSE}{SD^2} \quad (3.11)$$

where SD = observed standard deviation of item (or person) location measures on a Rasch scale, and MSE = mean of squared errors of item (or person) location measures. The reliability of separation index can range from 0 to 1. A higher value is always preferred indicating better separation of items (or persons) along a Rasch scale. It is worth noting that the reliability index does not reflect the quality of data. In a sense, it is recommended to use the residual statistics and fit indices to assess model-data fit. The reliability of separation provides additional information regarding the reproducibility of the scale that is also an important aspect in the scale evaluation.

3.3 Invariant Calibration of Items Across Subgroups of Persons

A requirement for achieving invariant measurement is to ensure invariant item calibration across subgroups of persons. The invariant calibration of items should be interpreted in three aspects. First, the rank ordering of items on the Wright map should be invariant across different subgroups of persons. Second, the estimated item difficulty for each subgroup should be invariant within the fluctuation range of sampling variability. Third, the items should measure an invariant construct across subgroups of persons.

Measurement invariance is an empirical matter that one should distinguish from the fundamental assumption of specific objectivity or invariant measurement (Andrich, 1988). Millsap (2011) defined measurement invariance as the measurement of a latent trait being the same across subpopulation groups. If an item functions differentially between subpopulation groups (e.g., female and male) at a given proficiency level, the measurement invariance is not achieved, and we call this differential item functioning.

Within the context of Rasch measurement theory, differential item functioning as statistical evidence is viewed as present when item difficulty estimates are significantly different across two or more subpopulation groups after adjusting for the overall scores of the respondents and placing the items on the same metric (Cohen & Kim, 1993). There is an important distinction between differential item functioning and item impact where the latter indicates actual difference in latent measures between two intact groups (Dorans & Holland, 1993).

Differential item functioning can hamper the reliability, validity, and fairness of the assessment system. It may cause systematic errors at a subgroup level that influence the reliability index. The ordering of items within each subgroup of persons may not be consistent, and the latent measures may not be objective. Penfield and Camilli (2006) indicated that assessing differential item functioning is critical for test validation of large-scale assessments, as well as an examination of selection bias and between-group differences on the underlying construct. The construct validity may be affected if an instrument also measures unintended construct (e.g., gender). Camilli (2013) discussed the concept of individual and group fairness being affected by differential item functioning. A comprehensive discussion on validity, reliability, and fairness issues is presented in Chapter 5.

In item response theory, unidimensionality is examined based on the principles of local independence. Local independence can be expressed as:

$$P(X|\theta, G = 1) = P(X|\theta, G = 2) = \cdots = P(X|\theta, G = k) \quad (3.12)$$

where X represents observed responses, θ denotes the measuring construct, and G shows the membership of k subgroups. The probability of obtaining a score of X for a person at a given proficiency level

44

should be invariant across subgroups. When an item presents differential item functioning, there may be another construct other than *G* affecting the probability of achieving a score (or endorsing an item), which violates local independence. Local independence is required to obtain the likelihood functions used by many estimation methods. Various approaches can be used to detect differential item functioning. First of all, we can visualize the differences in item difficulties between different subgroups through item response functions (Figure 3.3). An item response function displays the probability of endorsing an item (*y*-axis) against the ordered latent trait measures of persons (*x*-axis). Whether item response function curves coincide or not may imply differential item functioning. It is critical that a common scale (*x*-axis) needs to be established for respondents in different subgroups. Chapter 4 will discuss more approaches toward maintaining or establishing a common scale. Raju (1988) suggested calculating the area between two item response functions to examine differential item functioning. Mantel-Haenszel (MH; Mantel & Haenszel, 1959) is another commonly used approach to examine differential item functioning based on classical test theory, which examines the independence in the item scores between subgroups. Lord's (1977, 1980) Wald test is widely used for item response

Figure 3.3 Noncrossing Item Response Functions for Differential Item Functioning in Rasch Models

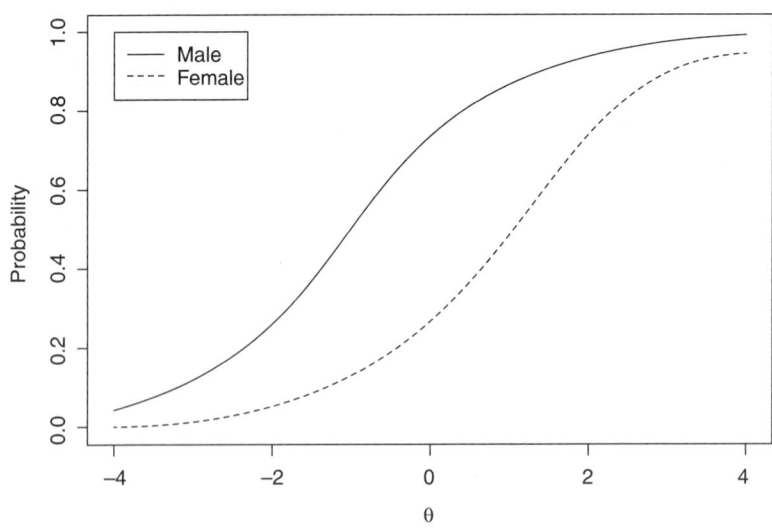

models to examine whether or not item parameter estimates differ between subgroups of persons.

For Rasch measurement models, the detection of differential item functioning can be defined as a two-stage process. The first step establishes a common scale across subgroups. The second step examines the difference in item difficulty measures between subpopulation groups using the Welch t-test. The Welch t-test is also known as Welch unequal variance t-test, named after Bernard Lewis Welch (Welch, 1947). It is derived based on an independent samples t-test that compare means between two populations. The Welch t-test is suggested for populations that have unequal variances and unbalanced sample sizes (Ruxton, 2006) since it is robust to the inflation of Type I error rate (Derrick, Toher, & White, 2016). To examine differential item functioning of each individual item for Rasch measurement models, the Welch t statistic is specified as below (Linacre, 2018b):

$$\text{Welch } t = \frac{b_1 - b_2}{\text{Joint } SE} \tag{3.13}$$

and

$$\text{Joint } SE = \sqrt{SE_1^2 + SE_2^2}$$

$$df = \frac{\left(SE_1^2 + SE_2^2\right)^2}{\frac{SE_1^4}{n_1 - 1} + \frac{SE_2^4}{n_2 - 1}}$$

where b_1 is the item difficulty measure and SE_1 is the standard error of the item difficulty measure in Group 1, and b_2 and SE_2 are for Group 2. Popular specialized Rasch software, such as the Winsteps and Facets, incorporated the Welch's t-test to examine differential item functioning.

3.4 Illustrative Analyses

This section illustrates the procedures for evaluating model-data fit with an empirical example based on FIE scale. In particular, the individual item and person misfit are assessed using residual statistics and fit indices (Infit and Outfit). The differential item functioning between subpopulations (female and male) is examined for each item on the FIE scale. The syntax for Facets analysis is included in the online Appendix (https://study.sagepub.com/researchmethods/qass/engelhard-rasch-models).

Table 3.1 Summary of Descriptive Statistics

Measures	Person	Item	Gender
M	0.00	0.00	−.80
SD	1.73	1.57	0.11
N	40	8	2
Infit			
M	0.99	0.97	0.94
SD	0.52	0.18	0.32
Outfit			
M	0.97	1.21	1.05
SD	1.44	1.35	0.82
Reliability of separation	0.64	0.91	0.00
χ^2 statistic	96.3*	68.7*	0.30
Degrees of freedom	39	7	1
Variance explained by Rasch measures		45.66%	

*$p < 0.01$.
M, mean; SD, standard deviation.
Source: Based on Coleman-Jensen, Rabbitt, Gregory, and Singh (2015).

Table 3.1 shows the location measures for both persons and items on a Rasch scale as well as a summary of fit indices. The item difficulty measures are centered at zero with a standard deviation of 1.57. The Infit mean square residuals for items have a mean value of 0.97 and a standard deviation of 0.18. The item Outfit mean square residuals center at 1.21 and have a standard deviation of 1.35. A large standard deviation indicates more variability in the Outfit mean square residuals across the items. The reliability of item separation is very high (0.91) implying high probability to reproduce the item location measures on the scale.

The person food insecurity measures center at zero to achieve the comparison between gender groups. The standard deviation of person latent measures on the scale is 1.73, which is slightly larger than the standard deviation of item location measures. This suggests comparable spread of the distributions of item and person location measures. The reliability of person separation is 0.64, and the chi-square test is significant, $\chi^2(39) = 96.3$, $p < 0.01$, indicating that the persons in our sample are well separated along the Rasch scale.

The difference in the food insecurity measures between female and male is -0.80, and it is not a significant difference according to the reliability of separation (0.00) and the chi-square test. An empirical Wright map is displayed in Figure 3.4 to show separate distributions for female and male. The Rasch measures in logits (log-odds) are displayed in the first column, followed by persons. The plus sign indicates a positive direction for person latent trait measures—higher Rasch location measure indicates more severe food insecurity. The location measures for female and male are separately displayed in the subsequent two columns. This display option helps the readers to visualize gender difference in food insecurity measures. The last column shows the item location measures with a minus sign indicating a negative direction—higher Rasch location measure implies more difficult-to-affirm.

We conducted residual analysis by dividing persons into six intervals (one logit for each interval) based on food insecurity measures. Table 3.2 shows observed proportion, model-based probabilities, along with raw and standardized residuals for each item. The items are ordered based on their difficulty measures. The mean is used for obtaining proportions and model-based probabilities for each interval.

The Infit and Outfit mean square errors are shown in Table 2.4. Based on the mean square values (MSE), the items can be defined using fit categories (Engelhard & Wind, 2018): (A) productive for measurement ($0.50 \leq MSE < 1.50$), (B) less productive for measurement but not distorting of measures ($MSE < 0.50$), (C) unproductive for measurement but not distorting of measures ($1.50 \leq MSE < 2.00$), and (D) unproductive for measurement and distorting of measures ($MSE \geq 2.00$). The Infit mean square errors of all eight items suggested the items are productive for measurement. Five out of eight items were also productive for measurement based on Outfit mean squares.

To provide a comprehensive report on the item evaluation, we created residual profiles that include Infit and Outfit mean square residuals, raw and standardized residuals for each interval of food insecurity measures, and residual plots that display standardized residuals against ordered person latent measures. The residual profiles for three selected items are shown here as an illustration. The fit categories for these three items are not A based on the Outfit mean square errors (Table 2.4). Item 3 (Table 3.3) has a large Outfit mean square which places this item in a fit category of D (i.e., unproductive for

48

Figure 3.4 The Wright Map

```
+-------------------------------------------------------------------------+
|Measr|+Person  |+Female    |+Male                         |-Items |
|-----+---------+-----------+------------------------------+-------|
|  3 +         +           |                              +       |
|     |         |           |                              | 8     |
|     |         |           |                              |       |
|     | *.      | F         | M  M                         |       |
|     |         |           |                              |       |
|     |         |           |                              |       |
|     |         |           |                              |       |
|  2 +         +           |                              +       |
|     |         |           |                              |       |
|     |         |           |                              |       |
|     |         |           |                              |       |
|     | **.     | F         | M  M  M  M                   |       |
|     |         |           |                              |       |
|     |         |           |                              |       |
|     |         |           |                              |       |
|  1 +         +           |                              +       |
|     |         |           |                              |       |
|     |         |           |                              | 7     |
|     | *       | F         | M                            |       |
|     |         |           |                              |       |
|     |         |           |                              | 1 4 6 |
|     |         |           |                              |       |
|  * 0 *  **    * F  F  F   | M                            *       *
|     |         |           |                              |       |
|     |         |           |                              | 5     |
|     |         |           |                              |       |
|     |         |           |                              |       |
|     | ***     | F  F      | M  M  M  M                   |       |
| -1 +         +           |                              +       |
|     |         |           |                              |       |
|     |         |           |                              |       |
|     |         |           |                              |       |
|     | ***     | F  F      | M  M  M  M                   |       |
|     |         |           |                              | 2     |
|     |         |           |                              |       |
| -2 +         +           |                              +       |
|     |         |           |                              |       |
|     |         |           |                              |       |
|     |         |           |                              | 3     |
|     |         |           |                              |       |
|     | ******* | F  F  F  F| M  M  M  M  M  M  M  M  M  M |       |
|     |         |           |                              |       |
| -3 +         +           +                              +       |
|-----+---------+-----------+------------------------------+-------|
|Measr| * = 2   |F = 1 female| M = 1 male                  |-Items |
+-------------------------------------------------------------------------+
```

Table 3.2 Residual Statistics for Intervals of Person Food Insecurity Measures

Group	Interval (logits)	Frequency	Item 3	Item 2	Item 5	Item 6	Item 4	Item 1	Item 7	Item 8
		Panel A: Observed proportion								
1	(2, 3)	3	0.67	1.00	1.00	1.00	1.00	0.67	1.00	0.67
2	(1, 2)	5	1.00	1.00	0.80	0.80	0.80	0.80	0.60	0.20
3	(0, 1)	2	1.00	1.00	0.50	0.50	1.00	1.00	0.00	0.00
4	(−1, 0)	10	0.90	0.50	0.60	0.30	0.40	0.40	0.30	0.00
5	(−2, −1)	6	0.67	0.67	0.17	0.33	0.00	0.00	0.17	0.00
6	(−3, −2)	14	0.50	0.36	0.07	0.00	0.00	0.07	0.00	0.00
		Panel B: Expected proportion								
1	(2, 3)	3	0.99	0.99	0.95	0.92	0.92	0.92	0.86	0.45
2	(1, 2)	5	0.98	0.96	0.86	0.78	0.78	0.78	0.66	0.20
3	(0, 1)	2	0.96	0.91	0.73	0.61	0.61	0.61	0.47	0.10
4	(−1, 0)	10	0.87	0.76	0.47	0.34	0.34	0.34	0.23	0.04
5	(−2, −1)	6	0.71	0.52	0.23	0.15	0.15	0.15	0.09	0.01
6	(−3, −2)	14	0.45	0.27	0.09	0.06	0.06	0.06	0.03	0.00
		Panel C: Raw residuals								
1	(2, 3)	3	−0.32	0.01	0.05	0.08	0.08	−0.25	0.14	0.22
2	(1, 2)	5	0.02	0.04	−0.06	0.02	0.02	0.02	−0.06	0.00

Table 3.2 *(Continued)*

Group	Interval (logits)	Frequency	Item 3	Item 2	Item 5	Item 6	Item 4	Item 1	Item 7	Item 8
3	(0, 1)	2	0.04	0.09	-0.23	-0.11	0.39	0.39	-0.47	-0.10
4	(-1, 0)	10	0.03	-0.26	0.13	-0.04	0.06	0.06	0.07	-0.04
5	(-2, -1)	6	-0.04	0.15	-0.06	0.18	-0.15	-0.15	0.08	-0.01
6	(-3, -2)	14	0.05	0.09	-0.02	-0.06	-0.06	0.01	-0.03	0.00
Panel D: Standardized residuals										
1	(2, 3)	3	-5.63	0.17	0.40	0.51	0.51	-1.62	0.70	0.77
2	(1, 2)	5	0.32	0.46	-0.39	0.11	0.11	0.11	-0.28	0.00
3	(0, 1)	2	0.29	0.44	-0.73	-0.32	1.13	1.13	-1.33	-0.47
4	(-1, 0)	10	0.30	-1.90	0.81	-0.29	0.37	0.37	0.54	-0.63
5	(-2, -1)	6	-0.22	0.72	-0.37	1.26	-1.03	-1.03	0.66	-0.25
6	(-3, -2)	14	0.38	0.73	-0.24	-0.95	-0.95	0.18	-0.66	0.00

Note. The items are ordered based on item difficulty measures.

Table 3.3 Residual Profile for Item 3 (Few Foods)

Item 3 (Infit MnSq: 1.12; Outfit MnSq: 4.50)

Group	Interval (logits)	Frequency	Expected proportion	Raw residuals	Standardized residuals
1	(2, 3)	3	0.99	−0.32	−5.63
2	(1, 2)	5	0.98	0.02	0.32
3	(0, 1)	2	0.96	0.04	0.29
4	(−1, 0)	10	0.87	0.03	0.30
5	(−2, −1)	6	0.71	−0.04	−0.22
6	(−3, −2)	14	0.45	0.05	0.38

MnSq, mean square fit statistic.

measurement and distorting of measures). Item 4 (Table 3.4) and Item 8 (Table 3.5) are in a fit category of B based on Outfit mean square (less productive for measurement but not distorting of measures). The residual plot for Item 3 has an obvious outlier. But for Item 4 and Item 8, standardized residuals are all within −2 and 2.

It is worthy of noting that Outfit indices detect outliers throughout the whole distribution.

In addition to the evaluation of item misfit, we should also examine person misfit. The goal of measurement is to provide reliable, valid, and

Table 3.4 Residual Profile for Item 4 (Skipped)

Item 4 (Infit MnSq: 0.69; Outfit MnSq: 0.47)				
Group	Interval (logits)	Frequency	Expected proportion	Raw residuals	Standardized residuals
1	(2, 3)	3	0.92	0.08	0.51
2	(1, 2)	5	0.78	0.02	0.11
3	(0, 1)	2	0.61	0.39	1.13
4	(−1, 0)	10	0.34	0.06	0.37
5	(−2, −1)	6	0.15	−0.15	−1.03
6	(−3, −2)	14	0.06	−0.06	−0.95

MnSq, mean square fit statistic.

fair scores to individuals. Therefore, we need to examine whether a person is a good fit to the constructed instrument. Edwards' (1948) error-counting method is a traditional method for examining the deviation between empirical-based and model-based scale (Table 3.6). To apply the error-counting method, we first ordered the items based on their difficulty measures from easiest to affirm to the hardest to affirm by columns. Furthermore, we rank ordered the persons based on food insecurity measures from most severe to least severe by rows. An ideal Guttman pattern assumes that a person who endorses a harder item

Table 3.5 Residual Profile for Item 8 (Whole Day)

Item 8 (Infit MnSq: 0.74; Outfit MnSq: 0.23)

Group	Interval (logits)	Frequency	Expected proportion	Raw residuals	Standardized residuals
1	(2, 3)	3	0.45	0.22	0.77
2	(1, 2)	5	0.20	0.00	0.00
3	(0, 1)	2	0.10	−0.10	−0.47
4	(−1, 0)	10	0.04	−0.04	−0.63
5	(−2, −1)	6	0.01	−0.01	−0.25
6	(−3, −2)	14	0.00	0.00	0.00

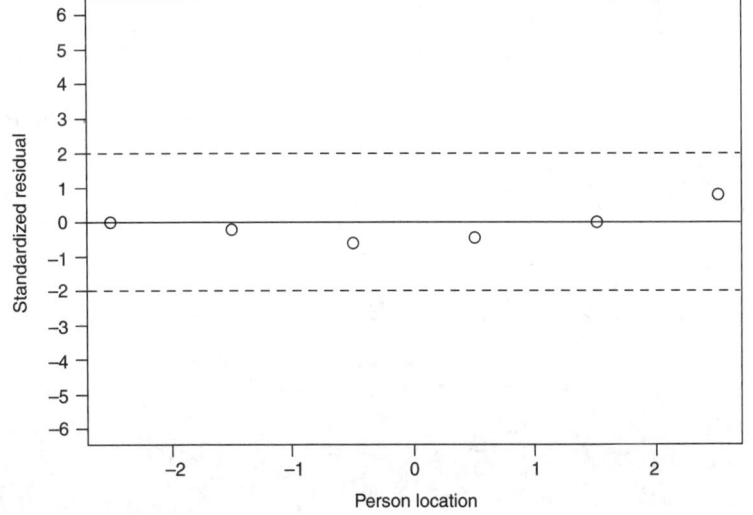

MnSq, mean square fit statistic.

would also endorse the easier ones. The gray-shaded cells should all be one underlying a model-based scale. The empirical responses that violate this assumption are viewed as errors which are bolded in Table 3.6. The total number of errors, denoted Guttman error, is used as an index to reflect person misfit. The number of Guttman errors ranges from 0 to 4. The person Infit and Outfit statistics displayed in Table 2.6 can be used as a combination to examine person misfit. Furthermore, we created the residual plots for persons based on standardized residuals across eight items.

Table 3.6 Reproduction of a Guttman Scale

		Item										Mean Squares		Fit Category	
PID	Measure	3 −2.38	2 −1.59	5 −0.32	6 0.21	4 0.21	1 0.21	7 0.81	8 2.86	Sum	Guttman Error*	Infit	Outfit	Infit	Outfit
22	2.63	1	1	1	1	1	1	1	0	7	0	0.36	0.16	B	B
29	2.63	1	1	1	1	1	0	1	1	7	2	1.8	1.61	C	C
23	2.63	0	1	1	1	1	1	1	1	7	2	2.02	9	D	D
13	1.47	1	1	1	1	1	1	0	0	6	0	0.59	0.41	A	B
37	1.47	1	1	1	1	1	1	0	0	6	0	0.59	0.41	A	B
17	1.47	1	1	1	1	1	0	1	0	6	2	0.81	0.64	A	A
33	1.47	1	1	1	1	0	1	1	0	6	2	0.81	0.64	A	A
25	1.47	1	1	0	0	1	1	1	1	6	4	2.02	1.83	D	C
35	0.67	1	1	1	0	1	1	0	0	5	2	0.72	0.54	A	A
26	0.67	1	1	0	1	1	1	0	0	5	2	0.88	0.72	A	A
7	−0.02	1	1	1	0	0	1	0	0	4	2	0.69	0.55	A	A
9	−0.02	1	1	0	1	1	0	0	0	4	2	0.87	0.68	A	A
21	−0.02	1	1	1	0	1	0	0	0	4	2	0.87	0.68	A	A
2	−0.02	1	0	0	1	1	0	1	0	4	4	1.4	1.35	A	B
24	−0.72	1	1	1	0	0	0	0	0	3	0	0.54	0.44	A	A
11	−0.72	1	1	0	1	0	0	0	0	3	2	0.72	0.61	A	A
1	−0.72	1	1	0	0	1	0	0	0	3	2	1.16	0.95	A	A
27	−0.72	1	0	1	0	1	0	0	0	3	2	1.16	0.95	A	A
28	−0.72	1	0	0	0	1	0	1	0	3	4	1.49	1.4	A	A

PID	Measure	1	2	3	4	5	6	7	8	Errors	Score	Infit	Outfit		
36	−0.72	**0**	**0**	**1**	0	1	0	0	1	4	3	2.13	2.13	D	D
39	−1.51	1	1	0	0	0	0	0	0	0	2	0.40	0.29	B	B
18	−1.51	1	1	0	0	0	0	0	0	0	2	0.40	0.29	B	B
20	−1.51	1	1	0	0	0	0	0	0	0	2	0.40	0.29	B	B
3	−1.51	1	**0**	**1**	0	0	0	0	0	2	2	0.92	0.68	A	A
12	−1.51	**0**	1	0	1	0	0	0	0	2	2	1.39	1.21	A	A
4	−1.51	**0**	**0**	0	1	0	0	**1**	0	4	2	2.16	2.49	D	D
6	−2.6	1	0	0	0	0	0	0	0	0	1	0.55	0.24	A	B
8	−2.6	1	0	0	0	0	0	0	0	0	1	0.55	0.24	A	B
10	−2.6	1	0	0	0	0	0	0	0	0	1	0.55	0.24	A	B
5	−2.6	1	0	0	0	0	0	0	0	0	1	0.55	0.24	A	B
16	−2.6	1	0	0	0	0	0	0	0	0	1	0.55	0.24	A	B
19	−2.6	1	0	0	0	0	0	0	0	0	1	0.55	0.24	A	B
40	−2.6	1	0	0	0	0	0	0	0	0	1	0.55	0.24	A	B
14	−2.6	**0**	1	0	0	0	0	0	0	2	1	1.04	0.48	A	B
15	−2.6	**0**	1	0	0	0	0	0	0	2	1	1.04	0.48	A	B
31	−2.6	**0**	1	0	0	0	0	0	0	2	1	1.04	0.48	A	B
32	−2.6	**0**	1	0	0	0	0	0	0	2	1	1.04	0.48	A	B
38	−2.6	**0**	1	0	0	0	0	0	0	2	1	1.04	0.48	A	B
30	−2.6	**0**	0	**1**	0	0	0	0	0	2	1	1.52	1.39	C	A
34	−2.6	**0**	0	0	0	0	**1**	0	0	2	1	1.62	2.25	C	D

*Guttman errors are highlighted in bold, and they are based on Edwards' (1948) error-counting approach. The items are ordered based on item difficulty measures.
PID, Person ID.

Figure 3.5 displays fit indices besides the residual plot for a few selected individuals that have large mean square errors. Person 23 has both large Outfit and Infit indices, and there is an outlier—the easiest item (Item 3) in the residual plot. Based on the response pattern, this person did not endorse the easiest item but affirmed the rest of the items. The Guttman error is 2 that occurred for Items 3 and 8. This is similar to Person 34. Since Guttman error happened to the items that are located in the middle, the Outfit indices were not ridiculously high. For Person 4 and Person 36, the variation of standardized residuals was large, and there were values outside the range between -2 and 2. These two persons have Guttman errors of 4.

For comparison purposes, the residual plots for those individuals who received an Infit or Outfit mean square lower than 0.5 were illustrated in Figure 3.6. In all the residual plots, the standardized residuals were quite close to zero with smaller variations. The response patterns had no Guttman error. Therefore, lower than 0.5 mean square errors implies no distortion of the Rasch measures, but a lack of variation across person measures may lead to less productive measurement.

Finally, we examined the differential item functioning between gender groups for each item. Table 3.7 lists the item difficulty measures in each gender group and the test statistics based on Welch t-test. In addition, Figure 3.7 displays the Welch t-values of differential item functioning analysis. The values to the left indicate higher endorsement by male group, and the values to the right indicate higher endorsement by female group. Based on the results, none of the items exhibited differential item functioning. However, Item 3 was quite close. Even though it did not reveal significant differential item functioning based on statistical evidence, content specialists should still examine the description of this item. Recall that Item 3 had a large Outfit mean square error (4.50) indicating unproductive for measurement and may even distort the measures. Model-data fit needs to be fulfilled before examining measurement invariance of items and persons.

3.5 Summary

This chapter focuses on the evaluation of model-data fit, especially the measurement invariance of items across subgroups of persons. Specifically, we discussed a measurement problem—differential item functioning. Model-data fit is a prerequisite for invariant measurement, which evaluates if empirical data fulfill the requirement of the proposed

Figure 3.5   Residual Plots for Person With Large Mean Square Errors

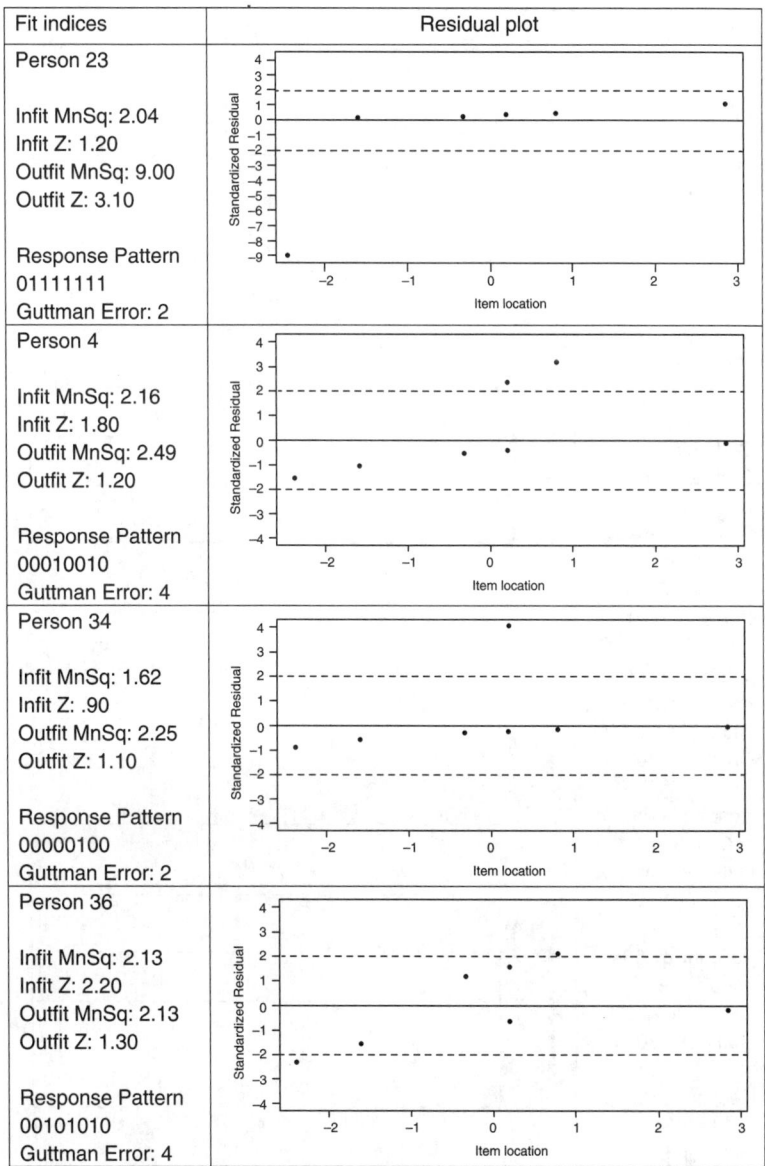

Note. For the response pattern, the items are ordered based on item difficulty measures, which are 3, 2, 5, 6, 4, 1, 7, 8.

58

Figure 3.6   Residual Plots for Person With Small Mean Square Errors

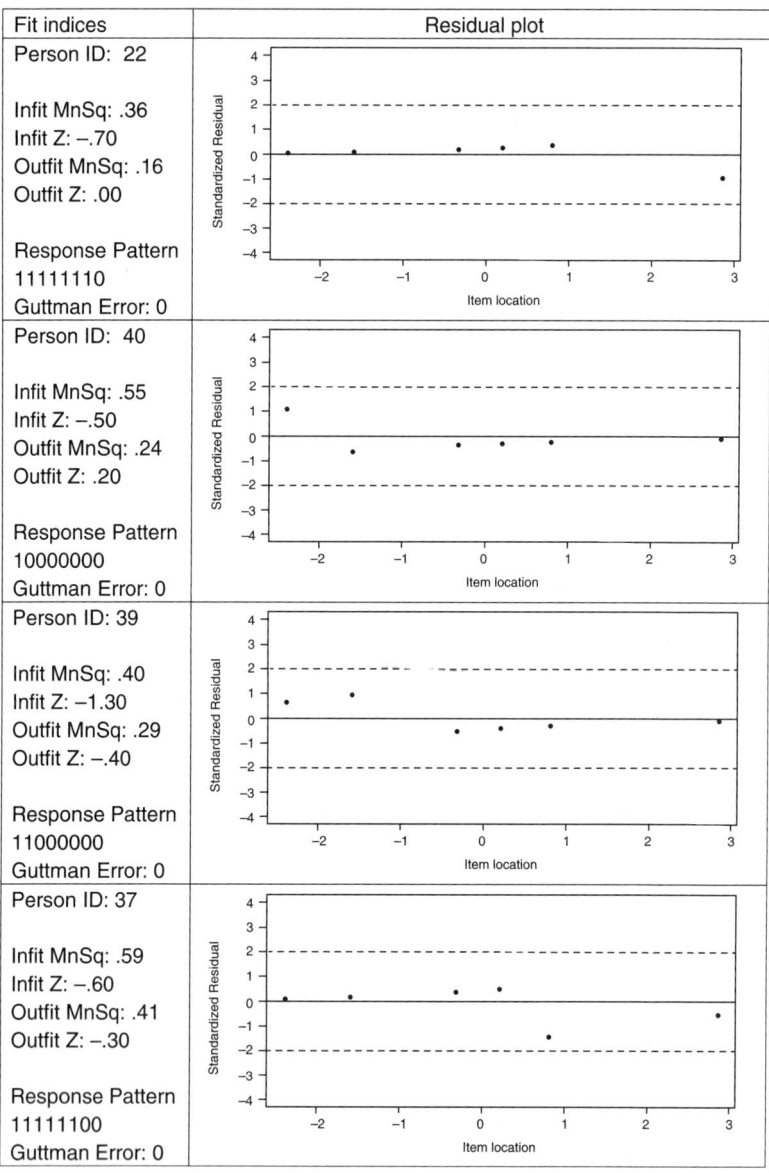

Fit indices	Residual plot
Person ID: 22 Infit MnSq: .36 Infit Z: −.70 Outfit MnSq: .16 Outfit Z: .00 Response Pattern 11111110 Guttman Error: 0	
Person ID: 40 Infit MnSq: .55 Infit Z: −.50 Outfit MnSq: .24 Outfit Z: .20 Response Pattern 10000000 Guttman Error: 0	
Person ID: 39 Infit MnSq: .40 Infit Z: −1.30 Outfit MnSq: .29 Outfit Z: −.40 Response Pattern 11000000 Guttman Error: 0	
Person ID: 37 Infit MnSq: .59 Infit Z: −.60 Outfit MnSq: .41 Outfit Z: −.30 Response Pattern 11111100 Guttman Error: 0	

Note. For the response pattern, the items are ordered based on item difficulty measures, which are 3, 2, 5, 6, 4, 1, 7, 8.

Table 3.7 Differential Item Functioning Across Gender Groups

	Male		Female			Joint	Welch		
Item	Measure	Square Error	Measure	Square Error	Contrast	Square Error	t	df	Probability
3	−1.57	0.49	−4.88	1.47	3.31	1.55	2.13	15	0.0501
1	0.64	0.59	−0.36	0.66	1.01	0.88	1.14	31	0.2630
6	0.64	0.59	−0.36	0.66	1.01	0.88	1.14	31	0.2630
2	−1.57	0.49	−1.66	0.67	0.09	0.83	0.11	26	0.9133
5	−0.83	0.51	0.57	0.72	−1.40	0.88	−1.60	25	0.1222
4	−0.29	0.53	1.13	0.78	−1.43	0.95	−1.50	24	0.1467
8	2.29	0.73	3.73	1.53	−1.44	1.70	−0.85	16	0.4079
7	0.31	0.57	1.84	0.90	−1.53	1.06	−1.43	23	0.1662

Figure 3.7 Interaction Plot Between Items and Gender Groups

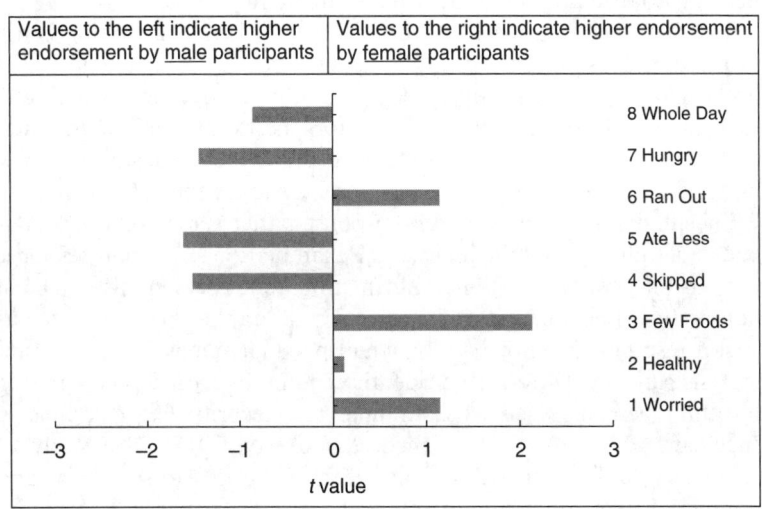

model (e.g., Rasch models) for defining the underlying continuum of the latent variable (Engelhard, 2009a). A good fitting model is based on the realization of model requirements.

Differential item functioning is a common measurement problem that challenges the measurement invariance of a Rasch scale. The procedures for assessing differential item functioning based on Rasch measurement theory are presented. Research in this area on differential item functioning studies has been rapidly developing in the past decades including differential item functioning for polytomous items (Potenza & Dorans, 1995), differential test functioning (Shealy & Stout, 1993), and differential rater functioning (Engelhard, 2008b). The major statistical approaches for detecting differential item functioning were discussed in Holland and Wainer (1993), such as SIBTEST (Shealy & Stout, 1993), logistic regression methods, standardization procedures (Dorans & Kulick, 1983, 1986; Dorans & Schmitt, 1991), and likelihood ratio test (Thissen, Steinberg, & Wainer, 1993). The book *Differential Item Functioning* (Holland & Wainer, 1993) is an edited volume that contributing authors are the experts on studying differential item functioning, and it covers an immense range of topics. In addition, Millsap (2011) and Osterlind and Everson (2009) are also good resources for researchers and practitioners to further study measurement invariance and differential item functioning.

Removing the items that exhibit differential item functioning is not always the best solution. The differences in item response functions that are found between demographic groups can be more meaningful through an exploration of other factors that differentiate the subpopulations. Further investigation of the construct-irrelevant variance may suggest modifications to items, tests, or test administrations.

Since individuals with different response patterns may obtain similar values of Infit and Outfit indices, an examination of person response patterns may reveal additional information about person misfit. The fit indices should be used with a combination of graphical displays such as person response functions and residual plots (Jennings, 2017). Cluster analysis can be used to further identify common characteristics among misfitting persons, such as examining test security (e.g., Perkins & Engelhard, 2013; Wollack, Cohen, & Eckerly, 2015). The study of person fit could be utilized to examine issues related to score bias and differential item functioning (Bolt & Johnson, 2009; Bolt & Newton, 2011).

We would like to emphasize that the invariant measurement concept should be discussed for the target population; however, subpopulation groups may vary in many ways, e.g., cultural differences. There is a clear need of having "one ruler for everyone" in social sciences, such as the FIE scale that is used to measure food insecurity globally (across countries and cultures). It is a challenging task to achieve invariant measures with a global scale. This not only relates to differential item functioning examining measurement invariance across countries but also is associated with the maintaining of the scale that will be discussed in Chapter 4 such as establishing a common scale to provide comparable measures across countries/regions.

Table 3.8 summarizes the indices and graphical displays for evaluating psychometric quality of items and objective measurement of persons based on Rasch measurement theory. We have discussed an empirical representation of latent scale—Wright map as well as the calibration of items and persons in Chapter 2. This chapter further presented fit indices for evaluating a Rasch scale along with residual

Table 3.8 Statistical Indices and Graphical Displays of Measurement Quality Based on Rasch Measurement Theory

Indices and Displays	1. Item Facet	2. Person Facet
1. Wright map	Where are the items located on the latent variable?	Where are the persons located on the latent variable?
2. Calibration and location of elements within facet	What is the location of each item (easy/hard)?	What is the location of each person (low/high)?
3. Standard errors for each element	How precisely has each item been calibrated?	How precisely has each person been calibrated?
4. Reliability of separation	How spread out are the item difficulties?	How spread out are the person locations?
5. Chi-square statistic	Are the overall differences between the items statistically significant?	Are the overall differences between the persons statistically significant?
6. Infit and Outfit statistics (mean square error)	Do the items fit the Rasch model?	Do the persons fit the Rasch model?
7. Unstandardized and standardized residuals	How different is each observed response from its expected value?	How different is each observed rating from its expected value?
8. Quality control charts, figures, and tables (displays of residuals)	What responses appear to be higher or lower than expected based on the Rasch model?	What responses appear to be higher or lower than expected based on the model?

Table 3.8 *(Continued)*

Indices and Displays	1. Item Facet	2. Person Facet
9. Differential item and person functioning	Are the item difficulties invariant across subgroups of persons?	Are the person locations invariant across subsets of items?
10. Unidimensionality (percentage of variance)	Can a unidimensional Rasch model be used to represent the data?	

plots as important graphical tools for visualizing the fit of individual item or person. The Infit and Outfit indices quantify the discrepancy between empirical and expected responses, and the fit categories are recommended for practical uses.

4

MAINTAINING A RASCH SCALE

This book considers four basic measurement problems: (1) defining a latent variable, (2) differential item functioning, (3) interchangeability of items, and (4) standard setting. This chapter focuses on the third issue related to the interchangeability of items. In this case, the measurement goal is to maintain the definition of a latent variable regardless of the particular items or scales used for measuring person locations on the latent variable. In this chapter, we address this measurement problem in terms of maintaining a Rasch scale.

A central goal in measurement is to define a latent variable that is represented by a unidimensional continuum. Rasch scales provide an approach for defining a latent variable or construct that can be visualized as a Wright map. It is important to recognize that the continuum representing the latent variable can be operationalized with different items, but that the representation of the underlying construct should remain stable and invariant. Lazarsfeld (1958) recognized the importance of invariant measurement. In his words:

> *In the formation of indices of broad social and psychological concepts, we typically select a relatively small number of items from a large number of possible ones suggested by the concept and its attendant imagery. It is one of the notable features of such indices that correlations with outside variables will usually be about the same, regardless of the specific "sampling" of items which goes into them from the broader group associated with the concept. This rather startling phenomenon has been labeled the "interchangeability of indices."*

(p. 113)

This quote highlights the importance of interchangeable items or scales that can yield comparable indices for person measurement. In this chapter, we discuss the importance of maintaining a Rasch scale in the social sciences with the goal of achieving invariant measurement of persons regardless of the particular items used to define a person's score. We continue to use the Food Insecurity Experience (FIE) scale for our illustrative analyses.

4.1 Comparable Scales for a Construct

An important goal for measurement is to maintain the definition of a latent variable regardless of the particular items or scales used for measuring person locations on the latent variable. Reliability, validity, and fairness are three fundamental psychometric qualities for evaluating the scores obtained from a measuring instrument. In addition to these three psychometric qualities, Messick (1994) explicitly mentions a fourth concept—*comparability*—as a basic psychometric quality that should be addressed for all measurement instruments. Moss (1992) also highlights a few specific measurement qualities that include comparability as one of the concerns regarding the consequences of score uses. The concept of comparability emphasizes the goal of obtaining comparable scores from different subsets of items or scales. Following Messick (1994), Mislevy (2018) defines comparability as the linking of different subsets of items or scales to generate a consistent definition of the latent variable. He highlights the importance of maintaining comparable scores. In a sense, interchangeable scales for a construct are desired to ensure persons measured by instruments consisting of different subsets of items can be compared. Messick (1995) emphasizes the importance of achieving comparable scales in measuring a latent variable. In his words:

Score comparability is clearly important for normative or accountability purposes whenever individuals or groups are being ranked. However, score comparability is also important even when individuals are not being directly compared, but are held to a common standard. Score comparability of some type is needed to sustain the claim that two individual performances in some sense meet the same local, regional, national, or international standard.

(p. 7)

The quest for obtaining comparable scales can be fulfilled by maintaining the Rasch scales given a prerequisite condition that all instruments measure the same construct. In other words, the ultimate goal of maintaining a scale is to provide comparable scores from different subsets of an items by establishing an invariant metric for the latent variable. Figure 4.1 depicts an illustration using two scales. First of all, we have a construct and the measurement goal is to obtain an estimate of the location of an object (θ_1) on this continuum. Next, we define the construct as height, and we want to measure the height of this object.

Figure 4.1 Measuring the Height of an Object (θ_1) With Two Scales

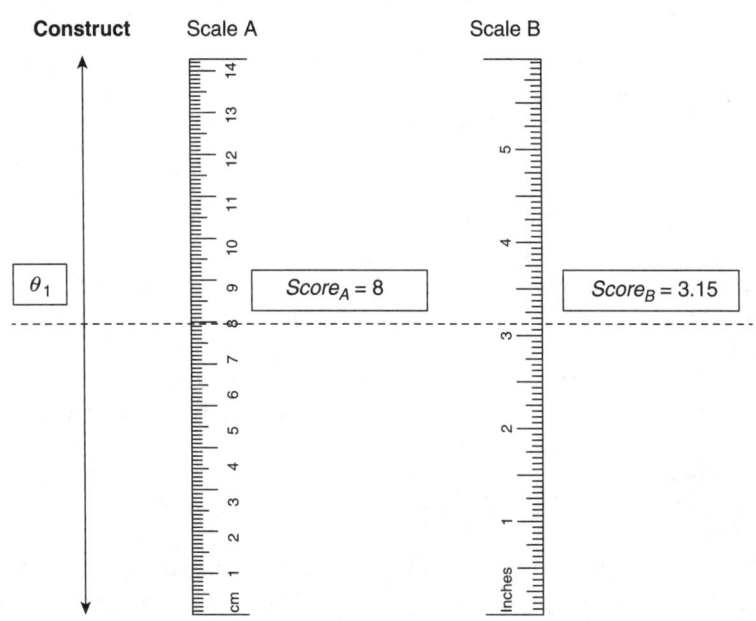

Scale A yields a score of 8, while Scale B yields a score of 3.15. However, we note that even though the two scales yield different scores, the *height* of the object does not change. Finally, we recognize that it is possible to convert Scale A into the same metric as used in Scale B (and vice versa) by linking the scales with a constant. Of course, the reader will recognize that the two scales are based on different metrics (centimeters and inches), and that a simple conversion formula can be used to obtain comparable scores that maintain their meaning across scales (1 cm = 0.394 inches and 1 inch = 2.54 cm). This issue also occurs for latent variables in the social sciences, but it is not typically recognized.

4.2 Invariant Measurement of Persons Across Scales

Once a scale has been developed, it is important to maintain the psychometric quality of the scale. A key aspect of this process is that the measurement of persons should be invariant over the particular items and even across different instantiations of the scale. It is important to recall that one of the requirements for invariant measurement according to Rasch (1961) measurement theory is that:

> *a comparison between two individuals should be independent of which particular stimuli [items] within the class considered were instrumental for the comparison.*
>
> (p. 331)

Therefore, the items should be interchangeable and can be viewed as a set of indicators from a larger item pool of potential items designed to measure the same construct. This chapter focuses on maintaining scales including items that can be used interchangeably to estimate item-invariant person measurement. For example, a scale developed to measure height and the units defined (e.g., inches, centimeters) can be utilized without concern about the particular ruler being used. The main point is that the underlying latent variable can be defined by different instruments and yet be maintained in such a way that the comparability of scores derived from different instruments is achieved.

The FIE scale discussed in this book is used internationally, and it is included as a part of the Gallup World Poll (http://www.gallup.com/poll/105226/world-poll-methodology.aspx). The FIE scale was created to define a global reference scale for measuring food insecurity (Cafiero, Viviani, & Nord, 2018). The actual items used in each country may vary because of differences in the language of administration, as well as the particulars of each country's social and cultural context. Maintaining a common metric for the FIE scale was accomplished by adjustments of each country's scale to match the global reference scale. These adjustments involve linking different scales as a methodology for obtaining comparable estimates of food insecurity across countries. Ballard, Kepple, and Cafiero (2013) have recognized the problem of creating scales with metric equivalence. In their words, the goal is to:

> *classify people into food insecurity classes in a way that is meaningful and comparable over time and across countries and socioeconomic contexts. Doing so requires: (a) establishing the metric equivalence of the scale, and (b) classifying cases into different food security levels taking into account possible differences in the severity of some items in some of the countries.*
>
> (p. 37)

In essence, Rasch measurement theory is used to establish metric equivalence by anchoring and scaling of the person parameters estimates obtained in different contexts or conditions to make sure that they are all expressed on a common scale.

Table 4.1 illustrates the underlying principle of linking scales. Food insecurity is defined by a global reference scale with eight items. As an illustration, imagine that two scales are constructed with different subsets of items that have been adapted to be appropriate for different cultures. Scale A is constructed by Items 1–5, while Scale B consists of Items 3–8. Suppose we know the food insecurity of a person defined by their location on the global reference scale is $\theta = -0.25$. If this person is administered Scale A, then it is expected that the person would respond Yes to items at or below the person's location (Items 1–4) with No responses to items located above the person's location (Item 5). Similarly, this person would also respond Yes to questions at or below Items 3–4 and No to questions above Items 5–8 in Scale B. Therefore, this person's expected score is 4 on Scale A and 2 on Scale B, while the person's location on the global reference scale remains the same ($\theta = -0.25$). The sum scores are dependent on the particular items and scales that were administered. Rasch measurement theory provides a methodology to adjust these sum scores obtained from two separate scales and produce invariant and comparable person scores.

Table 4.1 Illustration of Invariance Across Scales A and B for One Person (θ_1)

Global Reference Scale	Item	Label	Person	Scale A	Scale B
1.83	8	Whole Day			0
0.80	7	Hungry			0
0.45	6	Ran Out			0
0.41	5	Skipped		0	0
−0.25	4	Ate Less	$\theta_1 = -0.25$	1	1
−0.90	3	Healthy		1	1
−1.09	2	Few Foods		1	
−1.26	1	Worried		1	
			Sum Score	4	2

4.3 Illustrative Analyses

This section illustrates an approach that can be used to link two scales to obtain comparable measures on the underlying latent variable. The example is based on creating subsets of items from the data set used throughout this book. The edited data are shown in Table 4.2. There are 40 persons with the first 20 persons responding to Scale A that is defined by the five items from the original FIE scale (Items 1, 2, 3, 4, and 5). The second 20 persons responded to Scale B defined by six items from the original FIE scale (Items 3, 4, 5, 6, 7, and 8). Items 3, 4, and 5 are included in both scales as a common item link to aid in the development of a common metric. In educational testing, this is called the common item nonequivalent groups design. The person groups who responded to different scales are not required to have an equivalent distribution. The common items can link the two scales and establish a common metric.

The steps for linking the two scales with common items are based on the recommendations of Wright and Stone (1979). The details of this process are shown in Table 4.3. First, the items included in each scale are separately calibrated. Next, the difference in item locations on the common items (Items 3, 4, and 5) are calculated (Column 4, A − B). The mean difference in item locations is 0.85, and this is defined as the link used to obtain the locations of the items on an adjusted scale that is common across the two scales. The adjusted scale (column 7) is based on calibrating the two scales on a common reference scale. This process of common item linking (mean shift based on common items) is used extensively in educational testing.

A common scale can also be obtained simultaneously by calibrating the two scales with items not administered to one group given missing values (i.e., these values are missing by design). This procedure yields the total scale through a concurrent calibration based on all data responses. The last three columns in Table 4.3 show that these procedures yield similar estimates with small residual differences after standardizing the item locations between the adjusted Z scale and the total Z scale. Once an adjusted scale is developed to link forms, then comparable estimates of food insecurity can be obtained by using the Rasch measurement theory regardless of the particular items or scale that has been administered to different countries. The FIE scale was first calibrated separately by country and then was adjusted to develop a Global Reference Scale, but similar results can also be obtained with a simultaneous calibration.

Table 4.2 Two Food Insecurity Experience Scales

	Items							
Scale A	1	2	3	4	5			
Scale B			3	4	5	6	7	8
Person								
1	1	0	1	0	1			
2	0	0	1	1	1			
3	0	0	1	0	1			
4	0	0	0	0	0			
5	0	0	1	0	0			
6	0	0	1	0	0			
7	1	1	1	0	1			
8	0	0	1	0	0			
9	1	1	1	0	0			
10	0	0	1	0	0			
11	0	1	1	0	0			
12	0	1	0	0	0			
13	1	1	1	1	1			
14	0	1	0	0	0			
15	0	1	0	0	0			
16	0	0	1	0	0			
17	0	1	1	1	1			
18	0	1	1	0	0			
19	0	0	1	0	0			
20	0	1	1	0	0			
21			1	1	0	1	0	0
22			1	1	1	1	1	0
23			0	1	1	1	1	1
24			1	0	1	0	0	0
25			1	1	0	0	1	1
26			1	1	0	1	0	0
27			1	0	1	0	0	0
28			1	1	0	0	1	0
29			1	1	1	1	1	1

Table 4.2 *(Continued)*

					Items			
Scale A	*1*	*2*	*3*	*4*	*5*			
Scale B			*3*	*4*	*5*	*6*	*7*	*8*
Person								
30			0	0	1	0	0	0
31			0	0	0	0	0	0
32			0	0	0	0	0	0
33			1	0	1	1	1	0
34			0	0	0	0	0	0
35			1	1	1	0	0	0
36			0	1	1	0	1	0
37			1	1	1	1	0	0
38			0	0	0	0	0	0
39			1	0	0	0	0	0
40			1	0	0	0	0	0

4.4 Summary

This chapter describes issues related to the maintenance of comparable scales for defining a latent variable or construct. The specific focus is on establishing invariant measurement of persons across subsets of items. Essentially, the goal is to obtain metric invariance on an underlying continuum that does not depend on particular items or scales. The definition of the latent variable should be maintained regardless of the particular items and scales used for measuring person locations on the underlying latent variable.

There are several methods that can be used to maintain a comparable meaning of scores across scales with different subsets of items. In this chapter, we discuss two methods: common item linking and simultaneous calibration of scales. Common item linking involves identifying a set of items that are included in multiple scales, then the scales are linked. Linking scales to a reference scale can be accomplished either using the mean of the item locations of the common items (scales/ instruments) or by a simultaneous calibration. Simultaneous calibrations treat the items that are not administered as missing by design.

Table 4.3 Linking Two Scales With Common Items

Item	Scale A	Scale B	A − B	B + Link	(A + B + Link)/2	Adjusted Scale	Total Scale	Adjusted Z Scale	Total Z Scale	Residual
1	1.51					1.51	0.63	0.51	0.38	0.13
2	−1.02					−1.02	−1.60	−1.08	−0.95	−0.13
3	−3.15	−1.85	−1.30	−1.00	−2.08	−2.08	−2.96	−1.75	−1.76	0.02
4	2.16	−0.59	2.75	0.26	1.21	1.21	0.25	0.32	0.15	0.17
5	0.50	−0.59	1.09	0.26	0.38	0.38	−0.35	−0.20	−0.21	0.01
6		0.43		1.28		1.28	0.71	0.36	0.42	−0.06
7		0.43		1.28		1.28	0.71	0.36	0.42	−0.06
8		2.16		3.01		3.01	2.61	1.45	1.55	−0.10
Mean			0.85			0.70	0.00	0.00	0.00	0.00
SD						1.59	1.68	1.00	1.00	0.11

Note. The adjusted scale is established through a common set of items that define the link between the scales (link = 0.85), while the total scale is based on a simultaneous calibration of both scales. Z scales are standardized to have a mean of 0.00 and standard deviation of 1.00.

There are several reasons why researchers should maintain a Rasch scale to represent a latent variable. One reason is related to translation issues that arise when a particular set of items is not easily represented across cultures. Another reason is that when evaluating the impact of changing public policies, the researchers may not want persons responding to exactly the same set of items on each occasion. A third reason is that researchers may want to measure a common construct at extreme levels (e.g., low or high food insecurity) with targeted items that are still linked to the underlying continuum. Researchers in the natural sciences do not use the same instruments to measure the temperature of glaciers, persons, and volcanos. Nevertheless, they conceptualize temperature being meaningful along an extended continuum. A final context includes the use of a shorter version of a scale as a screener. We may want a shorter measure of food insecurity (e.g., three items) that can be used by pediatricians to identify risks of food insecurity. For all of these scenarios, common items should be available for linking the scales and producing comparable person scores.

As we build Rasch scales for key constructs, such as food insecurity, it is important to maintain scales over time and place among a community of scholars who are involved in research and policy related to the area of study. It is possible to have multiple comparable scales to measure one key construct. The latent variable is conceptually separate from the specific subsets of items that are included in a scale. For example, it is not necessary to use a particular ruler to measure height because rulers are viewed as interchangeable. The goal is to move the focus from particular items and scales to the construction of a stable set of constructs defined with invariant metrics that we can use to frame our understanding of the world. The method illustrated in this chapter is based on common items embedded across scales that can be linked to provide coherent indicators of the latent variable. The psychometric procedures used to maintain a latent continuum across subsets of items are called equating in educational testing. There are a variety of methods that can be used to link scales, but it is beyond the scope of this book to describe these methods. The interested reader should consult Kolen and Brennan (2004).

5

USING A RASCH SCALE

Once a Rasch scale has been created, it is important to consider a number of issues related to the use of the scale. Person scores obtained from a Rasch scale can be interpreted and used based on the *Standards for Educational and Psychological Testing* (*Test Standards*; AERA, APA, & NCME, 2014). The *Test Standards* are organized around three foundations of testing:

- Validity,
- Reliability, precision, and errors of measurement, and
- Fairness.

For each of these foundational areas, psychometric research focuses on the use of the person scores based on the intended purpose of the scale. The *Test Standards* reflect a consensus perspective in the measurement community, and there is still debate regarding how to evaluate the psychometric quality of a scale. Each of these areas is described in this chapter with attention to how the requirements of invariant measurement are related to the foundational areas. The foundations of the *Test Standards* related to the use of Rasch scales are also discussed. It should be noted that when scales are used for research purposes, the researcher still has an obligation to critically evaluate and regularly maintain the scale. The *Test Standards* can provide guidance on evaluating scale scores when they are used to guide research, theory, and policy.

Researchers create scales and use the person scores for a variety of purposes. One of the purposes is to inform policy, and a common policy use of the scales is to set cut scores or critical points on the continuum that define meaningful and ordered categories. Essentially, persons are placed in these categories based on judgments—this process is formally called standard setting (Cizek, 2012; Cizek & Bunch, 2007). There are a variety of procedures that can be used for setting cut scores to define meaningful categories on the continuum based on the Rasch scale. Examples in educational achievement include the Achievement Levels used in the National Assessment of Educational Progress (NAEP), such as Basic, Proficient, and Advanced (Hamm, Schulz, & Engelhard, 2011).

The Food Insecurity Experience (FIE) scale used in this book is typically reported in three categories (food secure, moderate food insecurity, and severe levels of food insecurity) for informing policy related to food insecurity. It is very important to critically evaluate whether or not the scale scores can be used to guide research, theory, and policy.

5.1 Three Foundations of Testing

This section describes three foundational areas of testing. Each of these foundational areas is discussed in terms of how they relate to the requirements of invariant measurement based on Rasch measurement theory, as well as the evaluative criteria set forth in the *Test Standards*. It is not widely recognized that the evidence suggested for supporting inferences about the meaning and interpretation of scale scores is to a large degree embedded within the measurement theories that are used to guide the development of the scale. As pointed out by Lazarsfeld (1966):

> *problems of concept formation, of meaning, and of measurement necessarily fuse into each other ... measurement, classification and concept formation in the behavioral sciences exhibit special difficulties. They can be met by a variety of procedures, and only a careful analysis of the procedure and its relation to alternative solutions can clarify the problem itself, which the procedure attempts to solve.*
>
> (p. 144)

Rasch models provide an approach for solving these measurement problems based on a coherent set of guidelines established using the principles of invariant measurement. The evaluation of the psychometric quality of a scale depends in a critical way on these three foundational areas of validity, reliability, and fairness.

Validity

The requirements of invariant measurement include the goals of developing item-invariant measurement of persons, person-invariant calibration of items, and a unidimensional scale representing the construct (latent variable).

The current consensus view of validity presented in the *Test Standards* (2014) is as follows:

> *Validity refers to the degree to which evidence and theory support the interpretations of test scores for proposed uses of tests. Validity*

is, therefore, the most fundamental consideration in developing and evaluating tests. The process of validation involves accumulating relevant evidence to provide a sound scientific basis for the proposed score interpretations.

(p. 11)

The essential point is that a "clear articulation of each intended test interpretation for a specified use should be set forth, and appropriate validity evidence in support of each intended interpretation should be provided" (*Test Standards*, 2014, p. 13). It should be noted that in this book we use Rasch measurement theory, and that the choice of measurement model plays an important role in determining the indicators that are used to evaluate evidence supporting the proposed meaning and use of the scores.

Table 5.1 lists the three clusters from the *Test Standards* used to describe evidence for validity. We include key questions related to each cluster and suggest indicators that can be used as sources of evidence for these aspects of validity.

Cluster I focuses on establishing the intended uses and interpretations of the scale scores. As shown in Table 5.1, we interpret this cluster as reflecting questions related to use, interpretation, and intended substantive meaning of the scale scores. In education, the intended uses typically include providing formative and summative information about student achievement in order to improve teaching and learning. For social science researchers, the focus tends to be more on the definition of a construct and the meaning of the scale scores within a broader theoretical framework. The FIE scale is designed to define the construct of food insecurity. The Wright map in Figure 1.6 represents the definition of the construct of food insecurity in terms of a set of items mapped onto a line.

Cluster II requires validity evidence related to the persons that are used in the development of the scale including a consideration of the settings. An important aspect of this cluster is to provide guidance on whether the scale is appropriate for the groups that the researcher plans to include in their study. It is important to recognize that although Rasch measurement theory provides the opportunity to achieve invariant measurement, the concept of specific objectivity used by Rasch stresses that the requirements must be examined for a specific data set. Invariance of scales is always dependent of examining model-data fit and a continuous evaluation of whether the scale is working as intended within a particular group of persons. For example, Engelhard,

Table 5.1 Key Questions for the Three Validity Clusters With Sources of Evidence

Cluster		Key Questions	Sources of Evidence
I. Establishing intended uses and interpretations		What is the construct (latent variable) being measured?	Definition of the construct
		What is the proposed substantive meaning of the scale scores including intended uses?	Description of the scale
			Wright map (intended/hypothesized)
II. Issues regarding samples and settings used in validation		What samples and settings are used to collect validity evidence to support the intended interpretations and uses of scale scores?	Detailed descriptions of the person samples and observation settings
III. Specific forms of validity evidence	A. Test content	Does the content of the items match the definition of the construct?	Expert judgment
			Content validity studies (alignment)
			Empirical item hierarchy
			Wright map (empirical)
	B. Response processes	How are responses categorized to represent person locations and levels on the Rasch scale?	Scoring rules
			Model-data fit (person fit)
			Person response functions
			Cognitive studies using qualitative methods
	C. Internal structure	Does the ordering of the items and persons conform to theoretical expectations based on hypothesized Rasch scale?	Wright map
			Model-data fit (items, persons)
			Evaluation of ordering of items and persons
			Unidimensionality
	D. Relations to other variables	Do the relationships between scale scores and other variables order the persons as expected based on a theoretical framework regarding the latent variable?	Program of research on the scale
			Differential item functioning across subgroups of persons
			Differential person functioning across subsets of items
			Criterion-related evidence
	E. Consequences of testing	What are potential intended and unintended consequences of the use of the scales scores?	Program of research on the scale

Figure 5.1 Plot of Item Locations (2012–2014, *y*-axis) on Anchored Item Locations (1998, *x*-axis) for the Household Food Security Survey Module

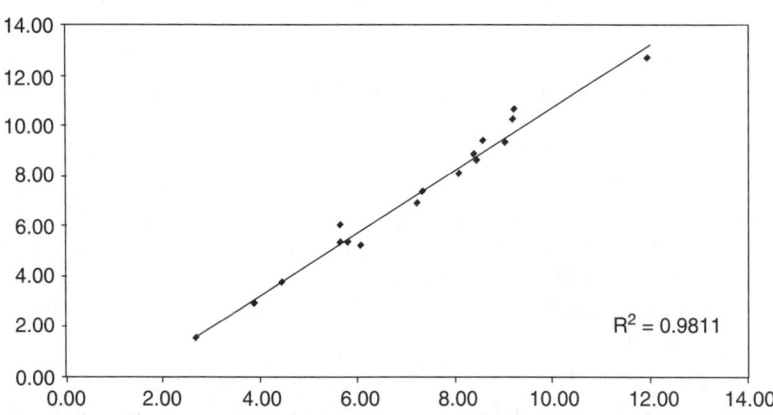

Note. Item locations are scaled to range between 0 and 14 logits with a mean of 7 on the Household Food Security Survey Module (Engelhard et al., 2016).
Source: Based on Coleman-Jensen, Rabbitt, Gregory, and Singh (2015).

Engelhard, and Rabbitt (2016) show how the invariance of the item locations on the Household Food Security Survey Module has been maintained over time (Figure 5.1).

The last cluster (Cluster III) includes five specific forms of validity evidence to support the interpretation and use of the scale:

A. Test content

B. Response processes

C. Internal structure

D. Relations to other variables

E. Consequences of testing

Each of these forms of validity evidence is described in turn.

Test content directly reflects how the latent variable is defined in terms of the items and tasks that provide the stimulus for person responses. There are several specific forms of evidence that can be used to evaluate whether the content of the items match the intended definition of the construct. First of all, expert judgments can be used to support the meaning of the items on an intended scale. Next, in educational settings, content validity studies may include alignment studies that are used frequently to document the quality of the items included in a scale. Another important source of evidence is the empirical Wright map that is obtained after administering the scale. The Wright map can be examined to evaluate whether the intended order of items (i.e., item hierarchy) is obtained. For example, the items on the FIE scale are ordered from easy to endorse (Item 2: Still thinking about the last 12 months, was there a time when you were unable to eat healthy and nutritious food because of a lack of money or other resources?) to hard to endorse (Item 8: During the last 12 months, was there a time when you went without eating for a whole day because of a lack of money or other resources?). This hierarchy of items can be used to provide empirical evidence regarding the meaning of scale scores on the latent variable.

Response processes refer to how responses are categorized on the scale. It also provides evidence of whether a person interacted as intended with the items on the scale; for example, in the case of rating scales (e.g., Likert scales), it is important to examine how persons used the scale. There are a variety of criteria for evaluating the functioning of rating scales (Engelhard & Wind, 2018). Person fit can also be used to determine if a person's response pattern is aberrant. Additionally, cognitive studies, such as think-aloud protocols, can be used to evaluate response processes.

Validity evidence related to the *internal structure* of a scale can be obtained in several ways. It examines whether the ordering of items and persons conforms to a hypothesized Rasch scale. The empirical Wright map shows the ordering of both persons and items on the scale. Researchers can examine the Wright map including model-data fit indices (item and person fit) to determine if evidence from the internal structure of the scale support the inference that scale is representing the latent variable as intended. The overall model-data fit and an examination of unidimensionality also provide information regarding the internal structure of the scale.

Evidence regarding *relations to other variables* is another important source of information about the scale. It is important to consider how the scores on the scale relate to other substantive variables. In many ways, this is essential for building up broader theoretical support for the scale. The functioning of the items, persons, and scales should also be examined across various subgroups of persons and subsets of items. Evidence of differential item functioning (measurement invariance) is frequently used for educational and psychological assessments. Differential person functioning is emerging as another source of evidence. Various types of criterion-related evidence can also be provided to support the validity of the scale.

Finally, it is important to consider the *consequences of testing*. Scales and person scores are developed to serve a specific purpose so that both intended and unintended consequences should be examined. In the policy context, the definition of latent variable (e.g., how we define food insecurity) can have profound effects on the policies developed to ameliorate the problem. For instance, based on an analysis of differential item functioning, Tanaka, Engelhard, and Rabbitt (2019) found that the households that receive Supplement Nutrition Assistance Program (SNAP) stamps report higher levels of food insecurity compared with non-SNAP participants who are at the same level of food insecurity. See Tanaka et al. (2019) for a detailed discussion of this apparent paradox related to SNAP and self-reports of food insecurity.

Scales created for research purposes may place a relatively higher emphasis on interpretation and meaning of the scale scores as compared with appropriate uses. One consequence is that the intended purpose in research studies may not require all of the evidence listed here, and it also suggests that the weight placed on each source of validity evidence may vary across purposes. Ultimately, we are developing warrants and support for persuading a community of practice including researchers regarding the meaning and use of the scale scores. Validity can be viewed as the invariance of meaning, interpretation, and use broadly conceived.

Reliability

Reliability broadly conceived includes examining the consistency of scale scores across different conditions. As pointed out in the *Test Standards* (2014), reliability refers "to the consistency or precision of scores across replications of a testing procedure" (p. 33). This perspective on reliability is strongly related to the concept of invariant measurement. In fact, examining consistency can be viewed as an evaluation of the hypothesis of invariant person ordering over different conditions, such as time and item subsets (test forms). Traditional views of reliability based on reliability coefficients focus on ordering of persons within groups, while the concept of precision relates to how much uncertainty there is in the estimation of each person's location on a scale. It should also be noted that the symmetry and duality of Rasch measurement theory provides a framework for extending the ideas regarding consistency and precision to item locations on a scale.

The *Test Standards* (2014) describes the reliability standards in terms of eight interrelated clusters. These are shown in Table 5.2. Each of these clusters is briefly described. These clusters can be grouped into three interrelated topics: research, indicators, and documentation. Clusters I, II, IV, and VII are related to a program of research on providing evidence regarding the consistency and precision of person measures over various types of replications for a scale. Clusters III, V,

Table 5.2 Key Questions for the Eight Reliability Clusters With Sources of Evidence

Cluster	Key Questions	Sources of Evidence
I. Specifications for replications (Research)	What aspects of the assessment system are invariant?	Program of research on the scale
II. Evaluating reliability and precision (Research)	Are there sources of error variance and uncertainty that can be identified?	Program of research on the scale
III. Reliability and generalizability coefficients (Indicators)	What indices of reliability and consistency are needed?	Reliability of separation for both persons and items
IV. Factors affecting reliability and precision (Research)	What types of variation are expected to influence the scale scores?	Program of research on the scale
V. Standard errors of measurement (Indicators)	What are the standard errors of measurement?	Standard errors of measurement for persons Standard errors of measurement for items
VI. Decision consistency (Indicators)	What is the level of consistency for policy classifications?	Indices of decision consistency
VII. Reliability and precision of group means (Research)	What aspects of the assessment system are invariant for group means?	Program of research on the scale
VIII. Documenting reliability and precision (Documentation)	What methods are used to report estimates of precision for scale scores?	Documenting the invariance of scale scores

and VI discuss the various indicators that can be used for evaluating the consistency and precision of scale scores. Finally, Cluster VIII provides guidance on documenting the evidence for other clusters—this information would be typically found in the technical manuals and other publications written to describe the psychometric quality of a scale.

The specification of the conditions for replication (Cluster I), the systematic evaluation of the sources of error variance and uncertainty (Clusters II and VIII), and a clear description of the various sources of measurement error that are expected to influence the precision of scale scores (Cluster IV) all reflect a broad program of research for evaluating the invariance of the scale scores. As pointed out by Simon (1990), "the fundamental goal of science is to find invariants" (p. 1). In order to evaluate how successful our scales are in achieving invariant measurement, researchers explore systematically whether the scale is functioning as intended over a variety of conditions. In other words, the researcher seeks aspects of the assessment situation that yield variant measures on a scale. For example, research has been conducted on the measurement invariance of the FIE scale across gender and countries (Wang, Tanaka, Engelhard, & Rabbitt, in press). This process of evaluating the consistency and invariance of person scores can be viewed as part of an ongoing program of research that critically evaluates the scale as an indicator of the latent variable.

Clusters III, V, and VI discuss standards for the specific psychometric indicators that can be used for reporting the consistency and precision of a scale over various replication conditions. In classical test theory, reliability coefficients are commonly used as statistical indicators of the consistency of scale scores for groups, while the standard error of measurement is used to represent the precision of the scale scores.

Rasch measurement theory provides evidence of consistency and precision with several interrelated indicators. The general concept of reliability defined as consistency and precision in scale scores can be viewed from the following four perspectives:

- Reliability of person scores (reliability of person separation on scale)

- Precision of person measures (standard errors of measurement for person locations on the scale)

- Reliability of item scores (reliability of item separation on scale)

- Precision of item calibrations (standard errors of measurement for item locations on the scale)

The reliability of person separation is comparable to coefficient alpha (Cronbach, 1951), while the standard error of measurement for persons reflects the precision of person scores. Lumsden (1977) provided a useful description of the concept of person reliability that highlights within-person variability. The reliability of item separation provides evidence of how well the items are spread out along the scale, while standard error of measurement for items reflects the precision of item calibrations. In other words, the reliability of item separation indicates the reproducibility of the Rasch scale with different person samples. If the person measures are categorized (e.g., low food insecurity, high food insecurity), then other indicators related to decision consistency indices can be used to provide useful information about the scale score usage.

As previously noted in terms of evidence regarding validity, the choice of measurement model is also relevant in considering the support for the inference that the scale scores reach an adequate level of reliability and precision.

In summary, issues related to reliability involve invariance of persons over replications. This concern goes back to Spearman (1904) who pointed out that "all knowledge—beyond that of bare isolated occurrence—deals with uniformities" (p. 72). Crocker and Algina (1986) echoed this concern: "Whenever a test is administered, the test user would like some assurance that the results could be replicated if the same individuals were tested again under similar circumstances. This desired consistency (or reproducibility) of test scores is called reliability" (p. 105).

Fairness

Fairness in testing is the last foundational area included in the *Test Standards*. According to the *Test Standards* (2014), "Fairness is a fundamental validity issue and requires attention throughout all stages of test development and use.... Fairness to all individuals in the intended population of test takers is an overriding, foundational concern" (p. 49). The guiding principle related to fairness is stated as follows:

All steps in the testing process, including test design, validation, development, administration, and scoring procedures, should be designed in such a manner as to minimize construct irrelevant variance and to promote valid score interpretations for the intended uses for all examinees in the intended population.

(*Test Standards*, 2014, p. 63)

83

Table 5.3 Key Questions for the Four Fairness Clusters With Sources of Evidence

Cluster	Key Questions	Sources of Evidence
I. Test design, development, administration, and scoring procedures that minimize barriers to valid score interpretations	What are potential sources of construct-irrelevant variance that may lead to inappropriate score interpretations for relevant subgroups?	Program of research on the scale Differential item functioning
II. Validity of test score interpretations for intended uses for the intended examinee population	What aspects of score interpretation and use require attention in order to ensure fairness to all individuals?	Differential person functioning
III. Accommodations to remove construct-irrelevant barriers and support valid interpretations of scale scores for their intended uses	What accommodations are needed to ensure the validity of individual interpretations of scores?	Program of research on the scale Identification of construct-irrelevant barriers
IV. Safeguards against inappropriate score interpretations for intended uses	What additional sources of information can be used to support the appropriate interpretations and uses of scores?	Program of research on the scale

Fairness is discussed in the *Test Standards* based on four interrelated clusters as described in Table 5.3.

The first cluster stresses the minimization of construct-irrelevant variance in all aspects of the scale construction process. Chapter 2 in this book discusses creating a Rasch scale, and researchers should consider issues of fairness related to each of the building blocks (Figure 2.1). The four building blocks for creating a Rasch scale are identification of a latent variable, development of an observational design that includes observable indicators (e.g., items), specification of scoring rules, and finally calibration of items with the Rasch model. In terms of the latent variable, the researcher should consider if the construct has the same meaning for everyone. For example, the concept of food insecurity may have a different meaning for persons in developed as compared with less developed regions of the world. It may also be the case that the particular items selected to define the latent variable have different meaning for subgroups of persons, such as males and females (Wang et al., in press). Scoring rules may also need to be adapted to reflect individual and subgroup variation. Finally, the Rasch model can be used to evaluate potential sources of construct-irrelevant variance through the examination of model-data fit including differential item and person functioning.

Both Clusters I and II stress the importance of ensuring fairness of interpretation and use for the intended persons to be measured. The *Test Standards* mention race, ethnicity, gender, age, socioeconomic status, linguistic, and cultural background as examples of relevant subgroups. Evidence regarding fairness for subgroups is typically provided through the use of differential item functioning, as described in Chapter 3, which includes an examination of measurement invariance (Millsap, 2011).

Fairness is frequently examined related to subgroups of persons, but the importance of individuals related to fairness issues has been succinctly stated by Wright (1984): "Bias found for groups is never uniformly present among members of the group or uniformly absent among those not in the group. For the analysis of item bias to do individuals any good ... it will have to be done at the individual level of the much more useful person fit analyses" (p. 285). Recent research has stressed including differential person functioning in the evaluation of the fairness of a scale (e.g., Engelhard, 2009a).

Clusters III and IV suggest a consideration of how to accommodate individuals who may not be able to interact with the measures as intended. Some of these accommodations are obvious, such as providing Braille forms of a scale for individuals who are visually impaired. Many of the accommodation issues involve a careful program of research on the scale to determine if the change in the assessments has altered the meaning of the scale scores and therefore affected the appropriate use of the scale scores.

In summary, fairness reflects invariance of the meaning of scores: "A test that is fair ... reflects the same construct(s) for all test takers, and scores from it have the same meaning for all individuals" (*Test Standards*, 2014, p. 50).

Even though the *Test Standards* address each foundational area separately, all of these areas function together. Some researchers have suggested that all of the criteria should be embedded within a broader view of validity (e.g., Messick, 1989), but a case could be made for considering the foundational area of fairness as the cornerstone of good measurement based on the principles of invariance.

In addition to these three key measurement concepts, Mislevy (2018) added a fourth concept: comparability. Following Messick (1994), he defines comparability as an extension of reliability and fairness issues to include a consideration of linking different sets of items or forms to generate comparable data and evidence. This topic is related to Chapter 4 addressing the general issue of maintaining a Rasch scale.

5.2 Standard Setting

Standard setting is a judgmental process used to determine cut scores on a scale. The standard-setting procedures identify significant categories that represent important milestones. For example, failing or passing a certification test to become a physician, failing or passing a high school graduation test, and being classified as food secure or insecure (Tanaka et al., 2019) are all examples of situations that require standard setting. In essence, standard setting can be viewed as a *values clarification process* based on the judgments of a carefully selected group of panelists (raters) who determine one or more cut scores on a continuum representing a latent variable. Standard setting involves an evaluative process that differs from a strict application of a set of psychometric rules—standard setting falls under the category of the use of Rasch scales for informing policy based on the definition of a continuum with critical cut scores that define meaningful and useful categories.

In educational measurement, formal research on procedures for standard setting was strongly motivated by the use of test scores in the 1970s for high stakes decisions such as high school graduation (Hambleton, Zenisky, & Popham, 2016). It reflected an important distinction between norm-referenced (NR) and criterion-referenced (CR) frameworks for attributing meaning to scale scores. In a nutshell, NR frameworks involve locating a person on a continuum based on a comparison to the norm group. NR scores are typically reported as percentiles. CR frameworks for score meaning focus on a comparison between a person's location on a continuum and a specific cut score or criterion (Wright, 1993). CR scores are typically reported in terms of ordered categories such as fail/pass. Similar approaches to standard setting have been adapted within the context of certification tests across a variety of professional occupations (Buckendahl & Davis-Becker, 2012).

Standard-setting procedures from a CR perspective can be broadly grouped into item-focused and person-focused methods. Item-focused methods ask panelists to rate items on whether or not a person above the cut score would pass or endorse the corresponding items. Person-focused methods are based on a comparable process by asking panelists if the person performances merit passing or not passing. Once a continuum is defined, then either process or a combination can be used as a specific standard-setting procedure.

Figure 5.2 illustrates the underlying principles for both methods. Item-focused methods would essentially ask panelists to review each of the items and then rate whether a minimally competent person (someone above the cut score) is expected to succeed on each item. If panelists determine that a minimally competent person would pass Items 1 and 2, but not Item 3, then the cut score would be set between Items 2 and 3 on the continuum. The use of person-based judgments is essentially the mirror image of this process with panelists rating whether

Figure 5.2　Conceptual Representation of Standard-Setting Methods Using Person-Based or Item-Based Judgments for Setting a Cut Score

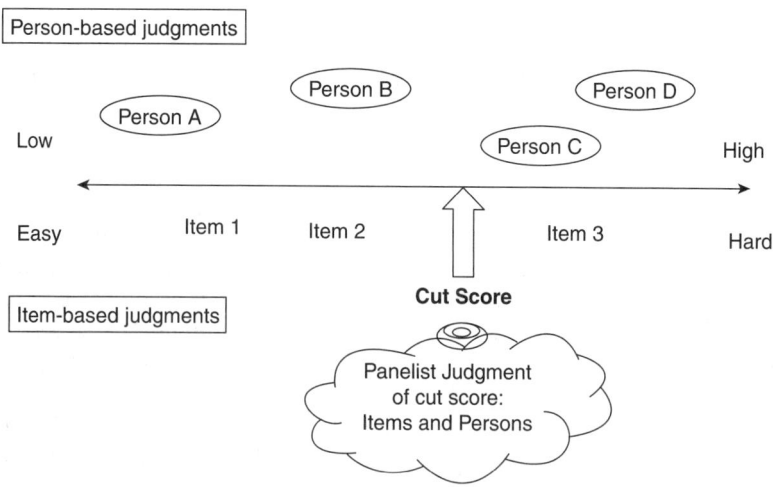

a person's performance on the assessment indicates success. If panelists identify Persons A and B as failing with Person C and D succeeding, then the cut score could be set between Persons B and C. Individual panelist judgments are collected in each general method and summarized to set an appropriate cut score on the continuum. Engelhard (2009b) has summarized a set of procedures for evaluating the judgments of standard-setting panelists based on Rasch measurement theory. We have selected four standard-setting methods that can be used to represent the basic distinctions between item-focused (Angoff and Bookmark) methods and person-focused (Contrasting Groups and Body of Work) methods.

Item-focused methods require panelists to make judgments about individual items. Most current methods have their genesis in a footnote from Angoff (1971):

[We] ask each judge to state the probability that the "minimally acceptable person" would answer each item correctly. In effect, judges would think of a number of minimally acceptable persons, instead of only one such person, and would estimate the proportion of minimally acceptable persons who would answer each item correctly. The sum of these probabilities would then represent the minimally acceptable score.

(p. 515)

There have been a variety of modifications to the Angoff method, but the focus remains clearly on panelists estimating where a *minimally acceptable person* would be located on a set of items that define a latent variable or line. Panelist judgments can be evaluated from a Rasch perspective (Engelhard & Anderson, 1998).

Another commonly used item-based method is the Bookmark method (Lewis, Mitzel, Mercado, & Schulz, 2012). A significant modification of the Bookmark method that warrants attention is the Mapmark method (Schulz & Mitzel, 2009). In this case, the item calibrations are presented to panelists with known locations (e.g., Wright map including only the items) that is called an ordered item booklet. Panelists determine the cut scores by identifying what items would be answered correctly and incorrectly by a borderline person (similar to the concept of Angoff's minimally acceptable person) and then placing a post-it note as a bookmark for the estimated cut score. Essentially, the panelist is creating an item response pattern for a borderline person. In some modified versions of the Bookmark method, panelists are asked to map performance-level descriptions to cut scores (Egan, Schneider, & Ferrara, 2012).

Panelist judgments based on the Bookmark method can also be evaluated with the Rasch model. Engelhard (2011) described the application of Rasch measurement theory to evaluate panelist judgments. The assessment was designed to measure mathematics proficiency among third graders in terms of four performance levels (Apprentice, Basic, Met, and Exceeded). There were 21 panelists with 57 items rated. The Wright map including the cut scores for the four performance levels with ratings collected over three rounds is shown in Figure 5.3. See Engelhard (2011) for additional information regarding the process.

Figure 5.3 Wright Map Based on Bookmark Judgments (Grade 3 Mathematics). A, apprentice; B, basic; M, met; E, exceeded.

```
|Logits +Items                                   +Panelists              +Rounds| R.1 | R.2 | R.3 |
+ 10 + 59                                         +                       +     + (E) + (E) + (E) +
|    |                                            |                       |     |     |     |     |
+  9 + 58                                         +                       +     +     +     +     +
|    |                                            |                       |     |     |     |     |
+  8 +                                            |                       +     +     +     +     +
|    |                                            |                       |     |     |     |     |
+  7 + 57                                         |                       +     +     +     + --- + CUT
|    |                                            |                       |     |     | --- |     |
+  6 + 56                                         |                       +     +     +     +     +
|    |   55                                       |                       |     | --- |     |     |
+  5 + 46  47  48  49  50  51  52  53  54 +       |                       +     +     +     +     +
|    |                                            |                       |     |     |     |     |
+  4 +                                            +                       +     +     +     +     +
|    |   44  45                                   |                       |     |     |     |     |
+  3 + 43                                         + 10                    +     +     + M   + M   +
|    |   38  39  40  41  42                       |                       |     | M   |     |     |
+  2 + 36  37                                     + 9                     +     +     +     +     +
|    |   34  35                                   | 8                     |     |     |     |     |
+  1 +                                            + 11  17  20  21  3     +     +     +     +     +
|    |   33                                       | 1   13  14  15        |     |     |     |     |
*  0 *                                            * 7                     *     * --- * --- *     *
|    |   31  32                                   | 19  5                 |     |     |     | --- | CUT
+ -1 + 28  29  30                                 + 18                    +     +     +     +     +
|    |   21  22  23  24  25  26  27               | 16   4   6            | 1   |     |     |     |
+ -2 + 16  17  18  19  20                         +                       + 2 3 +     +     +     +
|    |   15                                       |                       |     | B   |     |     |
+ -3 + 14                                         + 2                     +     +   B +     +     +
|    |                                            | 12                    |     |     | B   |     |
+ -4 +                                            +                       +     +     +     +     +
|    |                                            |                       |     |     |     |     |
+ -5 + 11  12  13                                 +                       + --- +     +     +     +
|    |                                            |                       |     |     |     |     |
+ -6 +                                            +                       +     + --- +     +     +
|    |                                            |                       |     |     | --- | CUT
+ -7 +                                            +                       +     +     +     +     +
|    |   10  9                                    |                       |     |     |     |     |
+ -8 + 4   5   6   7   8                          +                       +     +     +     +     +
|    |                                            |                       |     |     |     |     |
+ -9 +                                            +                       +     +     +     +     +
|    |                                            |                       |     |     |     |     |
+ -10 + 1   2   3                                 +                       +     + (A) + (A) + (A) +
```

Turning now to person-focused methods based on the judgments of panelists regarding performance of persons, we briefly discuss the Contrasting Groups method (Berk, 1976), and the Body-of-Work method (Kingston, Kahl, Sweeney, & Bay, 2001).

One of the earliest methods based on persons is the Contrasting Groups method (Berk, 1976). The Contrasting Groups method basically asks panelists to examine student work and then classify the student work into performance categories. The cut score is then set by finding the place where the groups have maximum separation.

The Body-of-Work method (Kingston et al., 2001) is a widely used method for setting standards based on person performances. The Body-of-Work method consists of several steps. First, panelists are trained to understand the purposes and goals of the standard-setting process. Next, the panelists engage in range-finding to identify variation in the quality of work samples being judged. Panelists provide ratings for Round 1 at this step. The third step is called pinpointing, and this step involves the selection and examination of additional work samples that focus more directly on preliminary performance standards obtained from range finding. Based on a consideration of the new work samples and summary ratings from Round 1, the panelists provide Round 2 ratings. The pinpointing step can be repeated with additional information provided to the panelists regarding the impact data. The ratings from the panelists define the categorization of the work samples into performance levels, such as below basic, basic, proficient, and advanced.

As was the case with panelist judgments based on the Bookmark method, panelist judgments from the Body-of-Work method can be evaluated with the Rasch model (Caines & Engelhard, 2009). The context was an alternative assessment of writing for students in high school. There were 18 panelists who rated 17 collections of evidence. Figure 5.4 presents the Wright map for the collections of evidence (COE) related to student work, the performance of the panelists over rounds (pinpointing and range finding), the location of the panelists, and finally the cut scores obtained for classifying the students.

In many policy settings outside of the contexts of educational assessment and certification testing, formal standard-setting processes have not been frequently used. For example, the FIE scale used for illustrations in this book for global monitoring is based on two specific items given the following process (Cafiero, Viviani, & Nord, 2018):

two thresholds [cut scores] have been set: one that identifies the level of severity beyond which a respondent would be classified as having

Figure 5.4 Wright Map for Collection of Evidence. B, below standard; COE, collection of evidence; M, meets standard; U, undecided (only used in range finding round).

```
+----------------------------------------------------------+
|Logit|+COE     |+Round       |+Panelists      |Scale|
|-----+---------+-------------+----------------+-----|
|   5 + 185 241 +             +                + (M) |
|     | 129     |             |                |     |
|     |         |             |                |     |
|     | 199     |             |                |     |
|     |         |             |                |     |
|   4 +         +             +                +     |
|     |         |             |                |     |
|     | 219     |             |                |     |
|     |         |             |                |     |
|     |         |             |                |     |
|   3 +         +             +                +     |
|     |         |             |                |     |
|     |         |             |                |     |
|     | 222     |             |                |     |
|     |         |             |                |     |
|   2 +         +             +                +     |
|     |         |             |                |     |
|     |         |             | 6              |     |
|     |         |             | 15             |     |
|     | 203     |             |                | --- | CUT
|   1 +         +             +                +     |
|     | 200     |             |                |     |
|     |         |             | 7   13 14      |     |
|     | 159 216 |             | 16             |     |
|     |         | Range Finding |              |     |
| *  0 *        *             * 1  3  5  8  10 * U * |
|     |         |             | 17 18          |     |
|     |         |             |                |     |
|     |         |             | 2   11         |     |
|     | 198     |             |                |     |
|  -1 +         +             +                +     |
|     | 179     | Pinpointing | 4   9          | --- | CUT
|     |         |             | 12             |     |
|     |         |             |                |     |
|  -2 + 148     +             +                +     |
|     |         |             |                |     |
|     | 136 186 |             |                |     |
|     |         |             |                |     |
|  -3 +         +             +                +     |
|     |         |             |                |     |
|     |         |             |                |     |
|     |         |             |                |     |
|  -4 +         +             +                +     |
|     | 167 172 |             |                |     |
|     |         |             |                |     |
|     |         |             |                |     |
|  -5 +         +             +                + (B) |
|-----+---------+-------------+----------------+-----|
```

moderate or severe food insecurity, and one that identifies severe levels only …. The first threshold is set to correspond to the severity of the 'ATELESS' item, while the second to the severity of the 'WHOLEDAY' item.

(p. 150)

The justification for these cut scores is as follows:

Individuals classified as having experienced moderate or severe food insecurity could be thus described as having eaten less than they thought they should at times during the year because they lacked sufficient resources for food, and most of them will have experienced more severe conditions. Those classified as having experienced severe food insecurity might have had high chances of going for whole days without eating, at times during the reference period, for the same reason of lack of sufficient resources to procure food.

(p. 150)

Figure 5.5 presents the Wright map for the FIE scale based on the data from Chapter 1.

It should be noted that there are 30 (75%) persons being classified as having low to moderate food insecurity and 10 (25%) persons classified as having moderate to severe food insecurity. None of the 40 persons exceeded the cut score based on the item corresponding to going a whole day (Whole Day) without food.

In summary, since standard setting involves panelists providing judgments, it can also be viewed as a rater-mediated process and being evaluated from that perspective (Engelhard, 2009b; Engelhard & Wind, 2018). As pointed out by Wright (1993):

We cannot set a standard unless we decide what we want, what is good, what is not, what is more, what is less … we must set up a line of increasing amounts, a variable which operationalizes what we are looking for … it is the calibration of the test's items which define the line.

(p. 1)

5.3 Summary

This chapter examined several issues related to using a Rasch scale. We organized the chapter around the three foundational areas—validity,

Figure 5.5 Wright Map for Food Insecurity Experience Scale With Cut Scores

```
|-------------------------------------------------------------------------------|
|        |                          |  Level of       |                         |
|Logit|+Person                      | Food Insecurity |-Items                   |
|-----+------------------------------+-----------------+-------------------------|
|      | High food insecurity        |                 | Hard to affirm          |
|  3 +  |                             +      Severe     +                         |
|      |                             |------- cut ------| Whole Day              |
|      | xxx                         |                 |                         |
|      |                             |                 |                         |
|      |                             |                 |                         |
|  2 +  |                             +                 +                         |
|      |                             |                 |                         |
|      |                             |                 |                         |
|      | xxxxx                       |                 |                         |
|      |                             |                 |                         |
|  1 +  |                             +                 +                         |
|      |                             |                 | Hungry                  |
|      | xx                          |                 |                         |
|      |                             |    Moderate     |                         |
|      |                             |------- cut ------| Ate Less   Ran Out   Worried |
*  0 *  | xxxx                       |       Low       *                         *
|      |                             |                 |                         |
|      |                             |                 | Skipped                 |
|      |                             |                 |                         |
|      | xxxxxx                      |                 |                         |
| -1 +  |                             +                 +                         |
|      |                             |                 |                         |
|      |                             |                 |                         |
|      | xxxxxx                      |                 | Few Foods               |
|      |                             |                 |                         |
| -2 +  |                             +                 +                         |
|      |                             |                 |                         |
|      |                             |                 | Healthy                 |
|      | xxxxxxxxxxxxxx              |                 |                         |
|      |                             |                 |                         |
| -3 +  |                             +                 +                         |
|      | Low food insecurity         |                 | Easy to affirm          |
|-----+------------------------------+-----------------+-------------------------|
```

Note. The full text for the eight items is described in Chapter 2.

reliability, and fairness that are included in the *Test Standards*. Evidence regarding validity, reliability, and fairness of a scale is used to evaluate the psychometric quality of a scale related to score interpretation and use. We present the discussion with an emphasis on how the concept of invariant measurement is related to each of the foundational areas. Rasch measurement theory provides a variety of indicators that can be used to evaluate the psychometric quality of a scale. Tables 5.1–5.3 summarize the three foundational areas, key questions, and sources of evidence related to each one.

An overview of standard-setting procedures is also provided because the creation of ordinal categories on a scale is commonly used to inform policy. Some standard-setting methods are based on Rasch measurement theory; for instance, the bookmark method depends on a well-calibrated Rasch scale with known item locations. We also illustrated the use of Rasch measurement theory for examining judgments of panelists in standard-setting processes. It is beyond the scope of this chapter to survey the plethora of methods and modified methods for setting cut scores. Cizek (2012) provides a good survey of major standard-setting methods.

6

CONCLUSION

The aim of science is to maximize the scope of solved empirical problems ....

–Laudan, 1977, p. 66

Beneath all of [Einstein's] theories, including relativity,

was a quest for invariants ... and the goal of science was to discover it.

–Isaacson, 2007, p. 3

Laudan (1977) has argued that progress in science is based on the successful resolution of empirical problems. Rasch measurement theory offers an approach for solving a variety of measurement problems. Many of the measurement problems can be viewed from the perspective of the pursuit of invariant measurement. In this book, we have focused on the solution of several empirical measurement problems including constructing a Rasch scale (defining a latent variable), evaluating a Rasch scale (examining measurement invariance and differential item functioning), maintaining a Rasch scale (equating, linking, and translating a scale), and using a Rasch scale (utilizing the foundational concepts of validity, reliability, and fairness from the *Test Standards* and the setting of performance standards).

We have grounded our view of Rasch measurement theory and the solution of measurement problems within the context of invariant measurement. We seek the construction, evaluation, maintenance, and use of Rasch scales that exhibit the desirable properties of invariant measurement. Invariant measurement yields the opportunity to create item-invariant person measurement and person-invariant item calibration when good model-data fit is obtained. We also stress that a lack of invariance can be of great interest. As pointed out by Nozick (2001):

What is objective about something, I have claimed, is what is invariant from different angles, across different perspectives, under different transformations. Yet often what is variant is what is especially interesting. We can take different perspectives on a thing (the more angles the better), and notice which of its features are objective and invariant, and also notice which of its features are subjective and variant.

(p. 102)

In the next section, we summarize our major points by chapter. This is followed by our reflections on future directions for practice, as well as research and theory related to Rasch measurement theory.

6.1 Key Themes by Chapter

There are a variety of measurement problems encountered in the social sciences. Many of these problems can be addressed as seeking invariant measurement with Rasch measurement theory as a useful approach. The book is organized around four components of scale development:

- constructing a Rasch scale,
- evaluating a Rasch scale,
- maintaining a Rasch scale, and
- using a Rasch scale.

In each chapter, we consider examples of measurement problems related to each of these components. This book considers in detail the following four measurement problems: definition of a latent variable, evaluation of differential item functioning, examination of interchangeability of items for person measurement (linking/equating), and creation of performance standards (cut scores) for standard setting.

In Chapter 1, we introduce the concept of invariant measurement. Three research traditions in measurement are described for organizing the plethora of measurement models available (test score, scaling, and structural traditions). Rasch measurement theory is an example of measurement models within the scaling tradition with a focus on creation of unidimensional Rasch scales that meet the requirements of invariant measurement—specifically, item-invariant person measurement and person-invariant item calibration.

Chapter 2 discusses the details of creating a Rasch scale. A major measurement problem considered in this chapter is how to define a latent variable. The key question is: What are the essential steps for constructing a Wright map based on Rasch measurement theory to empirically represent a latent variable? We answer this question by considering the four building blocks shown in Figure 2.1 (latent variable, observational design, scoring rules, and the Rasch model). We use the Food Insecurity Experience scale (Cafiero, Viviani, & Nord, 2018) to illustrate the construction of a Rasch scale.

Chapter 3 describes different approaches for evaluating the psychometric quality of a Rasch scale. The invariant measurement properties of Rasch measurement theory can be achieved only when there is good model-data fit. The model-data fit indices are based on the analyses of residuals that are summarized to examine item fit and person fit in detail. Differential item functioning is a potential measurement problem that explores factors that may distort the measures on a Rasch scale. Specifically, our example looks at model-data fit for the Food Insecurity Experience scale. The appropriate use and the meaning of scores are fundamentally dependent on achieving invariant measurement.

Chapter 4 introduces the next component—maintaining a Rasch scale for providing comparable person scores over different conditions. It is important to preserve the psychometric quality of a Rasch scale over a variety of conditions and to ensure the comparability of the Rasch measures. Many common measurement problems such as test equating, linking, as well as item parameter drift and scale drift relate to the maintenance of a scale. The interchangeability of items on a latent continuum is a very important feature of achieving the comparability of person scores. We discuss ways to maintain a Rasch scale given good model-data fit for a constructed scale. The ultimate goal is to establish a common scale for obtaining comparable person measures across a variety of conditions including different subsets of items. Equating and linking techniques reflect a selection of approaches that are relevant for solving this measurement problem.

Once a Rasch scale has been put into practice, it is important to consider issues related to the intended uses of the scale. Chapter 5 discusses how person scores obtained from a Rasch scale can be interpreted, evaluated, and used based on guidance from the *Test Standards* (AERA, APA, and NCME, 2014). According to the key themes in the *Test Standards*, we organized our discussion around three foundational areas: validity, reliability, and fairness. For policy purposes, we also discuss the use of Rasch measurement theory as a framework for setting performance standards in a statewide educational assessment and for setting cut scores on the Food Insecurity Experience scale that is used within the context of developing international policy on food insecurity.

6.2 Applications of Rasch Measurement Theory

In this section, we first consider some of the areas of practice that have utilized Rasch measurement theory, as well as areas of practice that may benefit from the creation and use of Rasch scales. Next, we briefly consider some of the research on extensions of Rasch models and Rasch measurement theory.

Figure 6.1 shows a steady increase in the frequency of citations on Rasch measurement theory in the 20th century. The total number of citations is 847 based on a Web of Science search using the topic phrase "Rasch Measurement Theory" covering the period of 1990 to September 9, 2019. The top five applications of Rasch measurement theory are related to Psychology (28%), Health Care Sciences and Services (15%), Educational Research (14%), Rehabilitation Sciences (9%), and Environmental and Occupational Health (9%). Aryadoust and Tan (2019) provide a systematic review on applications of Rasch measurement theory in psychology, medicine, and education.

The reader should consult Aryadoust and Tan (2019) for examples within various contexts that have successfully used Rasch measurement

Figure 6.1   Frequency of Citations With Theme of Rasch
           Measurement Theory E59

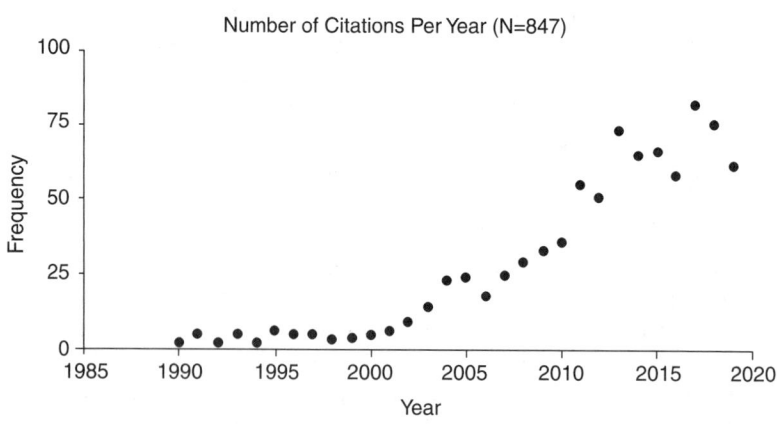

Source: Web of Science, September 2019.

Table 6.1 Selection of Key Books on Rasch Measurement Theory (1960–2019)

Decades	Authors	Titles
1960s	Rasch (1960/1980)	*Probabilistic models for some intelligence and attainment tests*
1970s	Wright and Stone (1979)	*Best test design: Rasch measurement*
1980s	Rasch and Wright (1980)	Rasch's book republished with foreword and afterword by Wright
	Wright and Masters (1982)	*Rating scale analysis: Rasch measurement*
	Andrich (1988)	*Rasch models for measurement*
	Linacre (1989)	*Many-facet Rasch measurement*
1990s	Wilson (1992, 1994). (Ed).	*Objective measurement: Theory into practice (Volumes 1–2)*
	Fischer and Molenaar (1995). (Eds).	*Rasch models: Foundations, recent developments, and applications*
	Engelhard and Wilson (1996). (Eds).	*Objective measurement: Theory into practice (Volume 3)*
	Wilson, Engelhard, and Draney (1997). (Eds.)	*Objective measurement: Theory into practice (Volume 4)*
2000s	Wilson and Engelhard (2000). (Eds).	*Objective measurement: Theory into practice (Volume 5)*
	Bond and Fox (2001)	*Applying the Rasch model: Fundamental measurement in the human sciences*
	Wilson (2005)	*Constructing measures: An item response modeling approach*
	Von Davier and Carstensen (2007). (Eds).	*Multivariate and mixture distribution Rasch models*
2010s	Garner, Engelhard, Wilson, and Fisher (2010). (Eds).	*Advances in Rasch measurement (Volume 1)*
	Brown, Drucker, Draney, and Wilson (2011). (Eds).	*Advances in Rasch measurement (Volume 2)*
	Engelhard (2013)	*Invariant measurement: Using Rasch models in the social, behavioral, and health sciences*
	Engelhard and Wind (2018)	*Invariant measurement with raters and rating scales: Rasch models for rater-mediated assessments*
	Andrich and Marais (2019)	*A course in Rasch measurement theory: Measuring in the educational, social and health sciences*

Note. This list should not be considered exhaustive—it is reflective of our personal journeys in understanding Rasch measurement theory.

theory to develop scales. Table 6.1 provides a selective list of books and edited volumes that include the research and practices of Rasch measurement theory.

As we move into the future, the forthcoming edition of *Educational Measurement,* 5th Edition (Cook & Pitoniak, in press), includes six application areas: (1) teaching and learning, (2) accountability in K-12 education, (3) admission, placement, and outcomes in higher education, (4) licensing and certification tests, (5) assessments of intrapersonal and interpersonal skills, and (6) international assessments. Rasch measurement theory can be used to address issues in each of these application areas. We briefly comment on each of the six areas in this section. The readers should recognize that each application area is connected to basic issues in measurement and assessment.

Rasch measurement theory can play an important role in classroom assessments to improve *teaching and learning.* One example is the BEAR assessment system developed by Mark Wilson and his colleagues at the University of California, Berkeley (Wilson, 2009). The use of four building blocks based on Rasch measurement theory leads to the creation of Rasch scale and a Wright map which plays a key role in conceptualizing and modeling learning progressions to improve classroom instruction and promote student learning.

Accountability is one of the pressing topics in American education. Accountability occurs at multiple levels with educational testing playing a role in the evaluation of students, teachers, and schools. Many of the measurement issues including the definition of the criteria used in these accountability systems depend in a fundamental way on the use of psychometrically sound scales. Accountability systems also increasingly use student growth and value-added models for evaluation purposes toward students, teachers, and schools in an education system. Rasch measurement theory can be used to support the validity, reliability, and fairness of scale scores for making decisions regarding current status and academic growth. In particular, performance standards in conjunction with Wright maps offer attractive systems for reporting accountability information at multiple levels in educational settings.

Assessments in higher education are important for making a variety of decisions including admission to colleges and universities, placement in classes, and accountability decisions. As with other applications based on scales, Rasch measurement theory can play an important role in supporting these decisions. In particular, Rasch scales and empirical Wright maps offer advantages over many of the current scales used to support decisions in higher education.

Another common use of tests is for *licensing and certification*. Rasch measurement theory offers psychometrically defensible frameworks for building licensing and certification systems. Rasch measurement theory is used extensively in these assessment systems. As with other areas that depend on reliable, valid, and fair scores to inform decisions, licensing and certification assessments can be built, evaluated, and maintained based on Rasch measurement theory.

Rasch measurement theory has also been used to develop scales representing a variety of affective variables. The use of these scales for measuring *intrapersonal and interpersonal skills* that are identified as important in education is a promising area of future research.

Another important application area relates to assessments that are designed to be used within *international contexts*. In the book, we look at a scale for measuring food insecurity that has been administered around the world (Cafiero et al., 2018). There are several major programs of international assessments in education settings such as the Program for International Student Assessment (PISA), the Trends in International Mathematics and Science Study (TIMSS), and the Progress in International Reading Literacy Study (PIRLS). These international assessments can be evaluated and used based on Rasch measurement theory to specifically explore the invariance of the scales and inform policy decisions in educational systems across cultures.

6.3 Concept Map for Rasch Models

This book focuses primarily on Rasch scales developed for dichotomous responses to a set of items. The reader should continue their studies of Rasch measurement theory and invariant measurement with various extensions of the Rasch models. Figure 6.2 summarizes a few specific and widely recognized members of the Rasch family of models: Dichotomous, Partial Credit, Rating Scale, Binomial Trials, Poisson Counts, and Facets. This family of Rasch measurement models is described in Wright and Masters (1982) with the Facets model added later by Linacre (1989).

Common extensions of Rasch models appear in several general frameworks: mixed models, multilevel models, and multidimensional models. First of all, mixed models offer a promising way of extending Rasch models by combining latent class analyses with Rasch models (Rost, 1990). This approach preserves the principles of invariant measurement by identifying subsets or classes of persons with good fit to a

Figure 6.2   Concept Map for Rasch Models

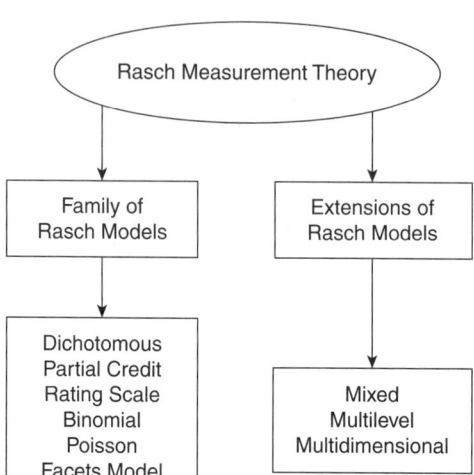

Rasch model. For example, Rost (2001) described a hierarchy of generalized Rasch models including mixed Rasch model (polytomous data), mixed linear logistic test model, mixed multidimensional model, and mixed multidimensional linear logistic model. Essentially, these extensions combine latent class analyses with a variety of Rasch models.

Multilevel models offer another exciting set of extensions to Rasch measurement theory (Adams, Wilson, & Wu, 1997; Kamata, 2001). The use of multilevel modeling framework accommodates nested data structures beyond person responses (e.g., schools, districts, and countries). The extension of Rasch measurement models to include higher levels provide a way to accommodate dependency due to nested structure of the response data with invariant measurement for persons as well as the examination of covariates across levels.

The final extension considered here relates to the development of multidimensional Rasch models. An approach is to view the Rasch model through the lens of multinomial logit models (Adams & Wilson, 1996; Adams, Wilson, & Wang, 1997). The use of multinomial logit models offers the flexibility to estimate different models based on various design matrices for persons and items. Adams and Wilson (1996) developed a unified approach for estimating the parameters of numerous Rasch models including the dichotomous, rating scale, partial credit, and facets models. Multidimensional Rasch models can also be approached from the perspective of structural equation modeling (Salzberger, 2011).

6.4 Final Words

One of our goals in this book is to describe a selection of measurement problems. Measurement issues are conceptual, and they can be obfuscated by complicated statistical machinery that may or may not be necessary—this was pointed out by Thorndike (1904) at the beginning of the 20th century, and the point is worth reiterating at the beginning of the 21st century. Measurement should not be a mysterious activity because it is fundamental for research, theory, and practice in the social, behavioral, and health sciences.

Rasch measurement theory can be used to create scales that reflect the perspectives of communities of practice who use the scales for a variety of purposes. We would like to stress that Rasch measurement theory leads to scales that represent what communities of practice want and expect from measures of key theoretical constructs. Measurement is an essential aspect of research and practice in the social, behavioral, and health sciences. One of the reasons for slow progress in these areas is that researchers are satisfied with scales that are inadequate to the tasks of supporting theories and applications. Progress depends on the creation of scales with meaningful and agreed-upon metrics for defining the major constructs in our substantive theories. Invariant measurement can be achieved through Rasch scales, and we encourage researchers to actively pursue the development of Rasch scales to support productive research, theory, and practice based on the four components we discuss in this book.

GLOSSARY

Classical test theory (CTT)—It is a theory of testing based on the idea that a person's observed score (O) on a test is the sum of a true score (T) and an error score (E)

Comparability—The linking of different subsets of items or forms to generate interchangeable indices

Differential item functioning—The examination of the conditional probabilities of success on an item between subgroups that have comparable locations on the latent variable

Differential person functioning—Method used for examining unexpected differences between the observed and expected performance of persons on a set of items

Errors of measurement—The difference between a measured quantity and its true value. It includes random error and systematic error

Fairness—Guidelines for developing and using assessment systems that lead to comparable measures of persons regardless of construct-irrelevant characteristics

Food Insecurity Experience (FIE) scale—The Food Insecurity Experience scale (FIES) is one of the four experience-based food insecurity scales which was developed by the Food and Agriculture Organization (FAO) through the Voices of the Hungry (VoH) project

Guttman error—Quantified discrepancy between observed and expected response patterns based on Guttman scaling (Edwards, 1948)

Guttman scaling—A technique for examining whether or not a set of items administered to a group of persons is unidimensional as proposed by Guttman

Infit mean square error—Weighted mean square error statistics that quantify the discrepancy between observed and expected response patterns based on residual analyses

Invariant measurement—A philosophical approach to measurement that supports item-invariant measurement of persons and person-invariant calibration of items that can be simultaneously represented by a variable map

Item-invariant person measurement—The measurement of persons should be independent of the particular items used to assign a location of a person on a latent variable

Item parameter/scale drift—The differential change in item parameter values over time

Item response function—A function that represents the relationship between person locations on the latent variable and probability of a positive response on an item

Item response theory (IRT)—A theory of testing based on the relationship between individuals' performances on a test item and the test takers' levels of performance on an overall measure of the ability that item was designed to measure

Latent variable—The underlying construct that is being measured or represented by Wright map

Metric equivalence—Two adjusted scales that measure the same construct are placed on a common continuum

Model-data fit—A judgment of the degree to which the data under investigation meet the requirements of a measurement model

Outfit mean square error—Unweighted mean square error statistics that quantify the discrepancy between observed and expected responses based on residual analyses

Person-invariant item calibration—The requirement that the location of items should be independent of the particular persons used to assign a location of an item on a latent variable

Person response function—The function that represents the relationship between response probabilities and a set of items for a person

Precision—A concept related to information matrix reflecting model precision. It is based on standard error of measurement

Reliability of separation—A value between 0.00 and 1.00 that indicates how well individual elements within a facet can be differentiated from one another

Residual—The difference between observed responses and model-based expectations

Scale linking—It is a process to place the item and person measures on a unidimensional continuum using a common metric

Scale reproducibility—It is the consistency of the constructed Rasch scale of the same latent variable across different person samples

Specific objectivity—Rasch (1960/1980) used this term to describe his version of invariant measurement. The concept is similar to objective measurement as used by Wright (1968)

Standard setting—It is a methodology used to define cut scores for categorizing levels of achievement or proficiency based on a latent scale

Standardized residual—It is defined as the residual divided by its standard deviation

Structural equation modeling (SEM)—It is a multivariate statistical analysis technique that is used to analyze structural relationship among observed and latent variables

Substantive theory—It is a theory developed for a substantive/empirical area to describe a formal/conceptual idea

Sum score—A score based on the summation of the observed responses (e.g., correct answers or affirmed statements)

Test equating—It refers to the statistical process of determining comparable scores on different test forms of the same assessment system

Three-parameter logistic (3PL) model—A response-centered measurement model expressed using a logistic item response function that includes three parameters for items (i.e., difficulty, discrimination, and guessing)

Two-parameter logistic (2PL) model—A response-centered measurement model expressed using a logistic item response function that includes two parameters for items (i.e., difficulty and discrimination)

Unidimensionality—A term used to describe a meaningful mapping of persons and items on to a single construct represented by a Wright map

Validity—An evaluative judgment based on evidence to examine whether or not an assessment system supports the intended meaning and use of scores

Venn diagram—It is a diagram that shows all possible logical relations between a finite collection of different sets

Vertical scaling—It refers to the process of placing test scores that measure the same construct but at different educational levels onto a common scale

Wright map—A visual representation of the construct being measured. Wright maps have also been called construct maps, item maps, curriculum maps, and variable maps

REFERENCES

Abelson, R. P. (1995). *Statistics as principled argument.* Hillsdale, NJ: Lawrence Erlbaum.

Adams, R. J., & Wilson, M. R. (1996). Formulating the Rasch model as a mixed coefficients multinomial model. In G. Engelhard, Jr. & M. R. Wilson (Eds.), *Objective measurement: Theory into practice* (Vol. 3, pp. 143–166). Norwood, NJ: Ablex.

Adams, R. J., Wilson, M., & Wang, W. (1997). The multidimensional random coefficients multinomial logit model. *Applied Psychological Measurement,* 21(1), 1–23.

Adams, R. J., Wilson, M. R., & Wu, M. L. (1997). Multilevel item response modelling: An approach to errors in variables regression. *Journal of Educational and Behavioral Statistics,* 22, 47–76.

American Educational Research Association, American Psychological Association, & National Council on Measurement in Education. (2014). *Standards for educational and psychological testing.* Washington, DC: AERA.

Andrich, D. A. (1985). An elaboration of Guttman scaling with Rasch models for measurement. In N. B. Tuma (Ed.), *Sociological methodology* (pp. 33–80). San Francisco, CA: Jossey-Bass.

Andrich, D. A. (1988). *Rasch models for measurement.* Thousand Oaks, CA: SAGE.

Andrich, D. A. (1989). Distinctions between assumptions and requirements in measurement in the social sciences. In J. A. Keats, R. Taft, R. A. Heasth, & S. H. Lovibond (Eds.), *Mathematical and theoretical systems* (pp. 7–16). North Holland: Elsevier Science.

Andrich, D. A. (2016). Rasch rating-scale model. In W. J. van der Linden (Ed.), *Handbook of item response theory, Vol. 1: Models* (pp. 75–94). Boca Raton, FL: CRC Press.

Andrich, D., & Marais, I. (2019). *A course in Rasch measurement theory: Measuring in the educational, social and health sciences.* Singapore: Springer Nature.

Angoff, W. H. (1971). Scales, norms, and equivalent scores. In R. L. Thorndike (Ed.), *Educational measurement* (2nd ed., pp. 508–600). Washington, DC: American Council of Education.

Aryadoust, V., Tan, H. A. H., & Ng, L. Y. (2019). A scientometric review of Rasch measurement: The rise and progress of a specialty. *Frontiers in Psychology*, 10, 2197.

Baker, F. B., & Kim, S. (2004). *Item response theory: Parameter estimation techniques* (2nd ed., revised and expanded). New York, NY: Marcel Dekker.

Baker, F. B., & Kim, S. (2017). *The basics of item response theory using R.* Cham: Springer.

Ballard, T. J., Kepple, A. W., & Cafiero, C. (2013). *The Food Insecurity Experience scale: Developing a global standard for monitoring hunger worldwide* (Technical Paper). Rome, Italy: FAO. Retrieved from http://www.fao.org/economic/ess/ess-fs/voices/en/

Berk, R. A. (1976). Determination of optimal cutting scores in criterion-referenced measurement. *Journal of Experimental Education*, 45, 4–9.

Bloom, B. S., Engelhart, M. D., Furst, E. J., Hill, W. H., & Krathwohl, D. R. (1956). Taxonomy of educational objectives: The classification of educational goals. In *Handbook 1: Cognitive domain*. New York, NY: David McKay.

Bollen, K. A. (1989). *Structural equations with latent variables*. New York, NY: Wiley.

Bolt, D. M., & Johnson, T. R. (2009). Addressing score bias and DIF due to individual differences in response style. *Applied Psychological Measurement*, 33(5), 335–352.

Bolt, D. M., & Newton, J. R. (2011). Multiscale measurement of extreme response style. *Educational and Psychological Measurement*, 71(5), 814–833.

Bond, T. G., & Fox, C. M. (2001). *Applying the Rasch model: Fundamental measurement in the human sciences*. Mahwah, NJ: Lawrence Erlbaum.

Brown, N. J., Drucker, B., Draney, K., & Wilson, M. (Eds.). (2011). *Advances in Rasch measurement* (Vol. 2). Maple Grove, MN: JAM Press.

Buckendahl, C. W., & Davis-Becker, S. L. (2012). Setting passing standards for credentialing programs. In G. J. Cizek (Ed.), *Setting performance standards: Concepts, methods, and perspectives* (pp. 485–501). Mahwah, NJ: Lawrence Erlbaum.

Cafiero, C., Viviani, S., & Nord, M. (2018). Food security measurement in a global context: The Food Insecurity Experience scale. *Measurement*, 116, 146–152.

Caines, J., & Engelhard, G. (2009). Evaluating body of work judgments of standard-setting panelists. Paper presented at the Annual Meeting of the American Educational Research Association, San Diego, CA.

Camilli, G. (2013). Ongoing issues in test fairness. *Educational Research and Evaluation*, 19(2–3), 104–120.

Cizek G. J. (Ed.). (2012). *Setting performance standards: Foundations, methods and innovation* (2nd ed.). New York, NY: Routledge.

Cizek, G. J., & Bunch, M. B. (2007). *Standard setting: A guide to establishing and evaluating performance standards on tests*. Thousand Oaks, CA: SAGE.

Cliff, N. (1983). Evaluating Guttman scales: Some old and new thoughts. In H. Wainer & S. Messick (Eds.), *Principles of modern psychological measurement: A Festschrift for Frederic M. Lord* (pp. 283–301). Hillsdale, NJ: Lawrence Erlbaum.

Cohen, A. S., & Kim, S. H. (1993). A comparison of Lord's χ^2 and Raju's area measures in detection of DIF. *Applied Psychological Measurement*, 17(1), 39–52.

Coleman-Jensen, A., Rabbitt, M., Gregory, C., & Singh, A. (2015). *Household food security in the United States in 2014* (Economic Research Rep. No. ERR-194). Washington, DC: U.S. Department of Agriculture, Economic Research Service.

Cook, L., & Pitoniak, M. J. (Eds.). (in press). *Educational measurement* (5th ed.).

Crocker, L., & Algina, J. (1986). *Introduction to classical and modern test theory*. New York, NY: Holt, Rinehart and Winston.

Cronbach, L. J. (1951). Coefficient alpha and the internal structure of tests. *Psychometrika*, 16, 297–334.

Cronbach, L. J., Gleser, G. C., Nanda, H., & Rajaratnam, N. (1972). *The dependability of behavioral measurements: Theory of generalizability for scores and profiles*. New York, NY: Wiley.

Derrick, B., Toher, D., & White, P. (2016). Why Welch's test is Type I error robust. *The Quantitative Methods in Psychology*, 12(1), 30–38.

Dorans, N. J., & Holland, P. W. (1993). DIF detection and description: Mantel-Haenszel and standardization. In P. W. Holland, & H. Wainer (Eds.), *Differential item functioning* (pp. 35–66). Hillsdale, NJ: Lawrence Erlbaum.

Dorans, N. J., & Kulick, E. (1983). *Assessing unexpected differential item performance of female candidates on SAT and TSWE forms administered in December, 1977: An application of the standardization approach* (ETS Research Rep. No. RR-83-9). Princeton, NJ: Educational Testing Service.

Dorans, N. J., & Kulick, E. (1986). Demonstrating the utility of the standardization approach to assessing unexpected differential item performance on the Scholastic Aptitude Test. *Journal of Educational Measurement*, 23(4), 355–368.

Dorans, N. J., & Schmitt, A. J. (1991). *Constructed response and differential item functioning: A pragmatic approach* (Research Rep. No. 91-47). Princeton, NJ: Educational Testing Service.

Edwards, A. L. (1948). On Guttman's scale analysis. *Educational and Psychological Measurement*, 8(3-1), 313–318.

Egan, K. L., Schneider, M. C., & Ferrara, S. (2012). Performance level descriptors: History, practice, and a proposed framework. In G. J. Cizek (Ed.), *Setting performance standards: Concepts, methods, and perspectives* (pp. 79–106). Mahwah, NJ: Lawrence Erlbaum.

Engelhard, G. (2005). Guttman scaling. In K. Kempf-Leonard (Ed.), *Encyclopedia of social measurement* (Vol. 2, pp. 167–174). San Diego, CA: Academic Press.

Engelhard, G. (2008a). Historical perspectives on invariant measurement: Guttman, Rasch, and Mokken [Focus article]. *Measurement: Interdisciplinary Research and Perspectives*, 6, 1–35.

Engelhard, G. (2008b). Differential rater functioning. *Rasch Measurement Transactions*, 21(3), 11–24.

Engelhard, G., Jr. (2009a). Using item response theory and model-data fit to conceptualize differential item and person functioning for students with disabilities. *Educational and Psychological Measurement*, 69(4), 585–602.

Engelhard, G. (2009b). Evaluating the judgments of standard-setting panelists using Rasch measurement theory. In E. V. Smith, Jr. & G. E. Stone (Eds.), *Criterion referenced testing: Practice analysis to score reporting using Rasch measurement models* (pp. 312–346). Maple Grove, MN: JAM Press.

Engelhard, G. (2011). Evaluating the bookmark judgments of standard-setting panelists. *Educational and Psychological Measurement*, 71(6), 909–924.

Engelhard, G. (2013). *Invariant measurement: Using Rasch models in the social, behavioral, and health sciences.* New York, NY: Routledge.

Engelhard, G., & Anderson, D. W. (1998). A binomial trials model for examining the ratings of standard-setting judges. *Applied Measurement in Education,* 11(3), 209–230.

Engelhard, G., Engelhard, E., & Rabbitt, M. P. (2016). Measurement of household food insecurity: Two decades of invariant measurement. *Rasch Measurement Transactions,* 30(3), 1598–1599.

Engelhard, G., Jr., & Perkins, A. F. (2011). Person response functions and the definition of units in the social sciences. *Measurement: Interdisciplinary Research and Perspectives,* 9, 40–45.

Engelhard, G., & Wilson, M. (Eds.). (1996). *Objective measurement: Theory into practice* (Vol. 3). Norwood, NJ: Ablex.

Engelhard, G., & Wind, S. A. (2018). *Invariant measurement with raters and rating scales: Rasch models for rater-mediated assessments.* New York, NY: Routledge.

Fischer, G. H., & Molenaar, I. W. (Eds.). (1995). *Rasch models: Foundations, recent developments, and applications.* New York, NY: Springer.

Garner, M., Engelhard, G., Wilson, M., & Fisher, W. (Eds.). (2010). *Advances in Rasch measurement* (Vol. 1). Maple Grove, MN: JAM Press.

Guttman, L. (1944). A basis for scaling qualitative data. *American Sociological Review,* 9(2), 139–150.

Guttman, L. L. (1947). On Festinger's evaluation of scale analysis. *Psychological Bulletin,* 44(5), 451.

Guttman, L. (1950). The basis for scalogram analysis. In S. A. Stouffer, L. Guttman, E. A. Suchman, P. F. Lazarsfeld, S. A. Star, & J. A. Clausen (Eds.), *Measurement and prediction* (Vol. IV, pp. 60–90). Princeton, NJ: Princeton University Press.

Hambleton, R. K., Zenisky, A. L., & Popham, W. J. (2016). Criterion-referenced testing: Advances over 40 years. In C. S. Wells & M. Faulker-Bond (Eds.), *Educational measurement: From foundations to future* (pp. 23–37). New York, NY: Guilford Press.

Hamm, C., Schulz, M., & Engelhard, G. (2011). Standard setting for the National Assessment of Educational Progress: Evidence regarding the transition from Angoff-based to Bookmark-based methods. *Educational Measurement: Issues and Practice,* 30(2), 3–14.

Holland, P. W., & Wainer, H. (1993). *Differential item functioning.* Hillsdale, NJ: Lawrence Erlbaum.

Isaacson, W. (2007). *Einstein: His life and universe.* New York, NY: Simon & Schuster.

Jennings, J. K. (2017). *A nonparametric method for assessing model-data fit in Rasch measurement theory* (Doctoral dissertation). University of Georgia, Athens.

Kamata, A. (2001). Item analysis by the hierarchical generalized linear model. *Journal of Educational Measurement, 38,* 79–93.

Karabatsos, G. (2000). A critique of Rasch residual fit statistics. *Journal of Applied Measurement, 1*(2), 152–176.

Kingston, N. M., Kahl, S. R., Sweeney, K., & Bay, L. (2001). Setting performance standards using the body of work method. In G. J. Cizek (Ed.), *Setting performance standards: Concepts, methods, and perspectives* (pp. 219–248). Mahwah, NJ: Lawrence Erlbaum.

Kolen, M. J., & Brennan, R. L. (2004). *Test equating, scaling, and linking: Methods and practices.* New York, NY: Springer.

Lane, S., Raymond, M. R., & Haladyna, T. M. (2016). *Handbook of test development* (2nd ed.). New York, NY: Routledge.

Laudan, L. (1977). *Progress and its problems: Toward a theory of scientific change.* Berkeley, CA: University of California Press.

Lazarsfeld, P. F. (1958). Evidence and inference in social research. In D. Lerner (Ed.), *Evidence and inference* (pp. 107–138). Glencoe, IL: Free Press.

Lazarsfeld, P. F. (1966). Concept formation and measurement in the behavioral sciences: Some historical observations. In G. J. Direnzo (Ed.), *Concepts, theory, and explanation in the behavioral sciences* (pp. 144–202). New York, NY: Random House.

Lewis, D. M., Mitzel, H. C., Mercado, R. L., & Schulz, E. M. (2012). The Bookmark standard setting procedure. In G. J. Cizek (Ed.), *Setting performance standards: Concepts, methods, and perspectives* (pp. 225–253). Mahwah, NJ: Lawrence Erlbaum.

Linacre, J. M. (1989). *Many-faceted Rasch measurement.* Chicago, IL: MESA Press.

Linacre, J. M. (2018a). *Facets computer program for many-facet Rasch measurement, version 3.81.0.* Beaverton, OR: Winsteps.com.

Linacre, J. M. (2018b). *Facets® Rasch measurement computer program user's guide*. Beaverton, OR: Winsteps.com.

Lord, F. M. (1977). A study of item bias, using item characteristic curve theory. In Y. H. Poortinga (Ed.), *Basic problems in cross-cultural psychology* (pp. 19–29). Amsterdam, the Netherlands: Swets & Zeitlinger.

Lord, F. M. (1980). *Applications of item response theory to practical testing problems*. Hillsdale, NJ: Lawrence Erlbaum.

Lumsden, J. (1957). A factorial approach to unidimensionality. *Australian Journal of Psychology*, 9(2), 105–111.

Lumsden, J. (1977). Person reliability. *Applied Psychological Measurement*, 1(4), 477–482.

Mantel, N., & Haenszel, W. (1959). Statistical aspects of the analysis of data from retrospective studies of disease. *Journal of the National Cancer Institute*, 22(4), 719–748.

Masters, G. N. (2016). Partial credit model. In W. J. van der Linden (Ed.), *Handbook of item response theory, Vol. 1: Models* (pp. 109–126). Boca Raton, FL: CRC Press.

McIver, J., & Carmines, E. G. (1981). *Unidimensional scaling* (Vol. 24). Thousand Oaks, CA: SAGE.

Messick, S. (1989). Meaning and values in test validation: The science and ethics of assessment. *Educational Researcher*, 18(2), 5–11.

Messick, S. (1994). The interplay of evidence and consequences in the validation of performance assessments. *Educational Researcher*, 23(2), 13–23.

Messick, S. (1995). Standards of validity and the validity of standards in performance assessment. *Educational Measurement: Issues and Practice*, 14(4), 5–8.

Millsap, R. E. (2011). *Statistical approaches to measurement invariance*. New York, NY: Routledge.

Mislevy, R. J. (2018). *Sociocognitive foundations of educational measurement*. New York, NY: Routledge.

Moss, P. A. (1992). Shifting conceptions of validity in educational measurement: Implications for performance assessment. *Review of Educational Research*, 62(3), 229–258.

Nozick, R. (2001). *Invariances: The structure of the objective world*. Cambridge, MA: Harvard University Press.

Osterlind, S. J. & Everson, H. T. (2009). *Differential item functioning* (Vol. 161). Thousand Oaks, CA: SAGE.

Penfield, R., & Camilli, G. (2006). Differential item functioning and item bias. In C. R. Rao & S. Sinharay (Eds.), *Handbook of statistics: Psychometrics* (Vol. 26, pp. 125–167). Amsterdam, the Netherlands: Elsevier.

Perkins, A. F., & Engelhard, G. (2013). Examining erasures in a large-scale assessment of mathematics and reading. Paper presented at the Annual Meeting of the American Educational Research Association, San Francisco.

Potenza, M. T., & Dorans, N. J. (1995). DIF assessment for polytomously scored items: A framework for classification and evaluation. *Applied Psychological Measurement*, 19(1), 23–37.

Raju, N. S. (1988). The area between two item characteristic curves. *Psychometrika*, 53(4), 495–502.

Rasch, G. (1960/1980). *Probabilistic models for some intelligence and attainment tests.* Copenhagen, Denmark: Danish Institute for Educational Research. (Expanded edition, Chicago, IL: University of Chicago Press).

Rasch, G. (1961). On general laws and meaning of measurement in psychology. In J. Neyman (Ed.), *Proceedings of the Fourth Berkeley Symposium on Mathematical Statistics and Probability* (pp. 321–333). Berkeley, CA: University of California Press.

Rasch, G. (1977). On specific objectivity: An attempt at formalizing the request for generality and validity of scientific statements. *Danish Yearbook of Philosophy*, 14, 58–94.

Raykov, T., & Marcoulides, G. A. (2011). *Introduction to psychometric theory.* New York, NY: Routledge.

Rost, J. (1990). Rasch models in latent class analysis: An integration of two approaches to item analysis. *Applied Psychological Measurement*, 14, 271–282.

Rost, J. (2001). The growing family of Rasch models. In A. Boomsa, M. A. J. van Duijn, & T. A. B. Snijders (Eds.), *Essays on item response theory* (pp. 25–42). New York, NY: Springer.

Ruxton, G. D. (2006). The unequal variance *t*-test is an underused alternative to Student's *t*-test and the Mann–Whitney *U* test. *Behavioral Ecology*, 17(4), 688–690.

Salzberger, T. (2011). *Specification of Rasch-based measures in structural equation modelling (SEM).* Retrieved from www.wu.ac.at/marketing/mbc/download/Rasch_SEM.pdf

Schulz, E. M., & Mitzel, H. (2009). A Mapmark method of standard setting as implemented for the National Assessment Governing Board. In E. V. Smith, Jr. & G. E. Stone (Eds.), *Criterion referenced testing: Practice analysis to score reporting using Rasch measurement models* (pp. 1–42). Maple Grove, MN: JAM Press.

Shealy, R. T., & Stout, W. F. (1993). A model-based standardization approach that separates true bias/DIF from group ability differences and detects test bias/DTF as well as item bias/DIF. *Psychometrika*, 58(2), 159–194.

Simon, H. A. (1990). Invariants of human behavior. *Annual Review of Psychology*, 41, 1–19.

Smith, R. M., & Hedges, L. V. (1982). A comparison of likelihood ratio χ^2 and Pearsonian χ^2 tests of fit in the Rasch model. *Educational Research and Perspectives*, 9, 44–54.

Spearman, C. (1904). "General intelligence," objectively determined and measured. *American Journal of Psychology*, 15, 201–293.

Stevens, S. S. (1951). Mathematics, measurement and psychophysics. In S. S. Stevens (Ed.), *Handbook of experimental psychology* (pp. 1–49). New York, NY: Wiley.

Stone, M. H., Wright, B. D., & Stenner, J. A. (1999). Mapping variables. *Journal of Outcome Measurement*, 3(4), 308–322.

Tanaka, V., Engelhard, G., & Rabbitt, M. P. (2019). Examining differential item functioning in the Household Food Insecurity Scale: Does participation in SNAP affect measurement invariance? *Journal of Applied Measurement*, 20(1), 100–111.

Thissen, D., Steinberg, L., & Wainer, H. (1993). Detection of differential item functioning using the parameters of item response models. In P. W. Holland & H. Wainer (Eds.), *Differential item functioning* (pp. 67–113). Hillsdale, NJ: Lawrence Erlbaum.

Thorndike, E. L. (1904). *An introduction to the theory of mental and social measurements*. New York, NY: Teachers College, Columbia University.

Thurstone, L. L. (1927). The method of paired comparisons for social values. *Journal of Abnormal and Social Psychology*, 21, 384–400.

Traub, R. (1997). Classical test theory in historical perspective. *Educational Measurement: Issues and Practice*, 16(10), 8–13.

van der LindenW. J. (Ed.). (2016). Preface. In *Handbook of item response theory, Vol. 2: Models* (pp. xviii–xix). Boca Raton, FL: CRC Press.

118

Von Davier, M., & Carstensen, C. H. (Eds.). (2007). *Multivariate and mixture distribution Rasch models*. New York, NY: Springer.

Wang, J., & Engelhard, G. (2019). Digital ITEMS module 10: Introduction to Rasch measurement theory. *Educational Measurement: Issues and Practice*, 38(4), 112–113.

Wang, J., Tanaka, V., Engelhard, G., & Rabbitt, M. P. (in press). An examination of measurement invariance using a multilevel explanatory Rasch model. *Measurement: Interdisciplinary Research and Perspectives*.

Welch, B. L. (1947). The generalization of Student's' problem when several different population variances are involved. *Biometrika*, 34(1/2), 28–35.

Wells, C. S., & Hambleton, R. K. (2016). Model fit with residual analyses. In W. J. van der Linden (Ed.), *Handbook of item response theory, Vol. 2: Models* (pp. 395–413). Boca Raton, FL: CRC Press.

Wilson, M. (Ed.). (1992). *Objective measurement: Theory into practice* (Vol. 1). Norwood, NJ: Ablex.

Wilson, M. (Ed.). (1994). *Objective measurement: Theory into practice* (Vol. 2). Norwood, NJ: Ablex.

Wilson, M. (2005). *Constructing measures: An item response modeling approach* (2nd ed.). Mahwah, NJ: Lawrence Erlbaum.

Wilson, M. (2009). Measuring progressions: Assessment structures underlying a learning progression. *Journal for Research in Science Teaching*, 46, 716–730.

Wilson, M. (2011). Some notes on the term: "Wright Map". *Rasch Measurement Transactions*, 25(3), 1331.

Wilson, M., & Engelhard, G. (Eds.). (2000). *Objective measurement: Theory into practice* (Vol. 5). Stamford, CT: Ablex.

Wilson, M., Engelhard, G., & Draney, K. (Eds.). (1997). *Objective measurement: Theory into practice* (Vol. 4). Norwood, NJ: Ablex.

Wilson, M., & Fisher, W. P. (Eds.). (2017). *Psychological and social measurement: The career and contributions of Benjamin D. Wright*. New York, NY: Springer.

Wollack, J. A., Cohen, A. S., & Eckerly, C. A. (2015). Detecting test tampering using item response theory. *Educational and Psychological Measurement*, 75(6), 931–953.

Wright, B. D. (1968). Sample-free test calibration and person measurement. In *Proceedings of the 1967 Invitational Conference on Testing Problems* (pp. 85–101). Princeton, NJ: Educational Testing Service.

Wright, B. D. (1980). Foreword and afterword. In G. Rasch (Ed.), *Probabilistic models for some intelligence and attainment tests* (pp. ix–xix, 185–196). Chicago, IL: University of Chicago Press.

Wright, B. D. (1984). Despair and hope for educational measurement. *Contemporary Education Review*, 3(1), 281–288.

Wright, B. D. (1993). *How to set standards*. Retrieved March 15, 2019, from https://www.rasch.org/memo77.pdf

Wright, B. D., & Masters, G. N. (1982). *Rating scale analysis: Rasch measurement*. Chicago, IL: MESA Press.

Wright, B. D., & Masters, G. N. (1984). The essential process in a family of measurement models. *Psychometrika*, 49, 529–544.

Wright, B. D., & Stone, M. H. (1979). *Best test design: Rasch measurement*. Chicago, IL: MESA Press.

INDEX